Who Governs the Internet?

Who Governs the Internet?

A Political Architecture

Robert J. Domanski

LEXINGTON BOOKS
Lanham • Boulder • New York • London

Published by Lexington Books
An imprint of The Rowman & Littlefield Publishing Group, Inc.
4501 Forbes Boulevard, Suite 200, Lanham, Maryland 20706
www.rowman.com

Unit A, Whitacre Mews, 26-34 Stannary Street, London SE11 4AB

British Library Cataloguing in Publication Information Available

Library of Congress Control Number: 2015949556
ISBN: 978-1-4985-1270-1 (cloth : alk. paper)
eISBN: 978-1-4985-1271-8

♾™ The paper used in this publication meets the minimum requirements of American National Standard for Information Sciences—Permanence of Paper for Printed Library Materials, ANSI/NISO Z39.48-1992.

Printed in the United States of America

This book is dedicated to my children, for whom, in everything I do, I only hope to make them proud. Now please let me sleep.

Contents

List of Illustrations

FIGURES

TABLES

Part I

INTRODUCTION & HISTORICAL CONTEXT

Chapter 1

Framing the Question, "Who Governs the Internet?"

As the Internet continues to become further integrated into all aspects of the global culture and economy, society has an increasing stake in pursuing socially beneficial and collective goals. Most people would agree, for instance, that society has a definite interest in preventing the dissemination of illicit child pornography or in mitigating the effects of widespread computer virus outbreaks. Some type of governance is vitally necessary to serve the interests of the public community, and indeed, such governance of the Internet has already emerged—although how these systems have emerged remains something of a puzzle. How have government institutions, private commercial firms, and the scientific academic community been able to create and implement rules and procedures for both the functional operation of the Internet and the behavior that takes place on it? To what extent and in what ways have these governance policies and arrangements emerged as a result of institutional decision-making and public policy processes at the federal level in the United States?

This book's main objectives will be, first, to develop a new model that deconstructs the Internet into four conceptual layers with the aim of helping scholars and policymakers better understand various Internet policy issues, and, second, to use this model in formulating a new political architecture that accurately maps out and depicts authority on the Internet by identifying who has decision-making authority and, therefore, a clear ability to shape behavior. We will then assess this four-layer model and its resulting map of political architecture by performing a detailed case study of U.S. national cybersecurity policy, post-9/11.

This study will examine the Internet from a public policy perspective, with a particular focus on policymaking processes and institutional arrangements. Specific institutions of various types have played a crucial historical role in shaping the direction of both how the Internet has evolved technologically as well as in setting the rules for how people use it. The Internet did not emerge spontaneously, nor did its present incarnation develop by accident. Rather, the Internet and all of its characteristics were consciously shaped as a direct result of explicit policy decisions.

The central question, then, is who governs the Internet? Which institutions, individuals, or other actors are shaping both the substance and direction of Internet governance policies? As the Internet continues to become more culturally and economically significant, it is important to investigate what type of governance is emerging and why it is emerging in that way.

WHAT DO WE MEAN BY "GOVERNANCE"?

Let's begin by dispelling a common myth. If you ask a random person off the street today, "Who governs the Internet?," the reply is typically, "Nobody does." Such is the prevailing wisdom—originating out of early Internet folklore that mistakenly equated decentralization with ungovernability. However, to the individual replying that nobody governs the Internet, a follow-up question along the lines of, "So can you do anything you want on the Internet without fear of consequence?," the reply is also just as frequently, "Of course not." And there's the disconnect. Many people still stubbornly hold onto the notion that the Internet cannot be governed, and yet nearly all of those same people freely admit that others have authority over their actions.

My contentions are that the Internet is, in fact, being governed; that it is being governed by specific and identifiable networks of policy actors; and that an argument can be made as to how it is being governed.

For one undeniable example that the Internet is being governed, take the existence of its DNS, or domain name system. The reason why when a user types the Uniform Resource Locator (URL) "www.google.com" into their web browser they can reliably expect to reach the website of the Google Search engine is that a public-private hybrid institution named ICANN (the Internet Corporation for Assigned Names and Numbers) has been developed over time to create a system for administering Internet domain names, ensuring their uniqueness, linking them to IP addresses, creating requirements for registration, and implementing formal dispute resolution mechanisms. ICANN is a nonprofit institution that was originally created by private-sector actors in response to a mandate issued by the U.S. Department of Commerce under the Clinton administration, which sought to cede control over the management of the Internet's system of centrally coordinated identifiers, for reasons which will be explored in Chapter 2. ICANN manages the DNS system, maintaining the Internet's operational functionality, using a multi-stakeholder model that incorporates businesses, governments, civil society organizations, and academic and scientific organizations, and is international in scope.[1] It is responsible for overseeing the Internet's core root name servers and all of its 639 generic top-level domains (gTLDs), including the core six—.com, .org, .net, .edu, .gov, and .mil,—and 248 country code top-level domains (ccTLDs),[2] as well as for making decisions over the adoption of future TLDs, which have sometimes proven to be controversial.[3] The very fact that the DNS exists and keeps the Internet operational is direct evidence of governance policy, and certainly refutes notions of the Internet being "ungovernable."

So what do we mean, then, in asking, "Who governs"? In academia, the phrase "Internet Governance" is fairly well established. Milton Mueller, a scholar of political economy, was an early researcher in the field and adopted a narrow definition, referring to the phrase only in terms of the Internet's functional operation with research focusing on the ICANN and DNS systems.[4] To this day, the phrase "Internet Governance" largely is used only in reference to the administration of the DNS system. Subsequent scholars like Laura DeNardis have similarly adopted this narrow definition in examining the adoption of the Internet's technical protocols like IPv6 and the roles played by a small handful of semipublic international consortium groups like the Internet Engineering Task Force (IETF) in standards-setting processes.[5]

But why should use of the phrase "Internet Governance" be so limited? I argue that a far broader, more comprehensive definition is called for. It ought to include the study not only of who makes decisions over the Internet's technical functionality, but also of who holds authority over much of the activity that takes place on it. In other words, we also want to be able to study the follow-up question of, "Can you do anything you want on the Internet without fear of consequence?" and "Why or why not?"

So, when asking, "who governs," the definition which will be used here adopts a broad policymaking approach and views governance as having three criteria: (1) the ability to constrain behavior; (2) the ability to enable behavior; and (3) the ability to produce intentional effects. Actors are said to govern when they have clear decision-making authority to create and implement policies with intentional effects that meet all three of these criteria.

To be clear, the issue here is one of governance, not government.[6] From Robert Dahl to C. Wright Mills, scholars have long sought to determine who has power, why they have it, and how they use it. In pluralist theory, power has many dimensions and is held in varying degrees by numerous actors—from individual people to large corporations to formal governmental institutions. Indeed, Dahl's approach in famously asking "Who Governs?" was to question how various interest groups compete in the political sphere, and that governance is ultimately determined by the relative capacities of different actors to influence governmental decision-making.[7] The questions at hand, in the context of the Internet, remain how all of those different actors are organized in creating and exercising their relative levels of authority. However, what sets the Internet apart from Dahl's analysis, as will be demonstrated time and again, is that on the Internet, it is not merely a matter of government having final decision-making authority, but also, to a considerable degree, numerous private actors as well. The Internet governance dynamic is characterized by various competing interest groups not only trying to influence government, but also competing to influence each other, and sometimes government trying to influence them. Identifying who holds authority versus who is trying to wield influence, perhaps more clear in Dahl's day, is an increasingly difficult task. Thus, not only do we need to ask who has power, but also who has more power than whom?

Understanding who has the power to govern is a political question. Governance is inextricably linked with concepts of power, and in that context, both must be defined for the purpose of this project. This is not to say that we plan on comprehensively defining these two ideas at the heart of Political Science—governance and power—once-and-for-all. Rather, it is necessary to clearly state which definitions will be used to carry out our specific research.

The literature on governance has markedly shifted in recent years from focusing on hierarchical governmental structures toward greater reliance on horizontal, hybridized, and associational forms of governance.[8] In the field of public administration, for instance, scholars such as Frederickson and Smith have observed this re-focus from the bureaucratic state and direct government to the "hollow state" and "third-party government."[9] Governance theories that incorporate ideas about the role of "conjunctions" or "associations" among organizational entities have become increasingly widespread.[10]

This "governance fever"[11] focusing on horizontal relationships between public- and private-sector actors has seen a deconstruction of the governance concept into several categorical types. *Network governance*, most frequently used for characterizing the Internet, is commonly associated with ideas of "self-governance" or "self-regulation." It refers to loosely structured coordination among numerous actors that function like an "organic or informal social system."[12] Network governance, as Taylor has argued, arises because of modern societies' complexities and their consequent requirement for distributed knowledge acquisition and decentralized problem-solving.[13] In contrast, *hierarchical governance* embraces the activities of government, law, and statutory regulation.[14] It describes processes that are characterized by vertical integration and managerial control within a set of lead institutions, and is the traditional method of analysis for studying top-down bureaucratic organizations. Meanwhile, *market governance* is equated with the forces of effective free-market competition with the invisible hand governing behavior.[15] There have also recently been new additional theories developed as scholars have sought to meaningfully depict what's occurring on the Internet specifically. *Adhocratic governance*, for example, is based on the idea of policy being made "ad-hoc," meaning in an improvised, on-the-fly type of manner, and that decision-making is guided by simply dealing with problems as they arise.[16] According to scholars like Mintzberg, "adhocracy" is a system superior to bureaucracy and one that will even eventually replace it. It is "any form of organization that cuts across normal bureaucratic lines to capture opportunities, solve problems, and get results."[17]

These are some of the theories about governance, but when it comes to actually defining the broader concept of the term, specifically from a policymaking perspective, the approach undertaken by Lawrence Lessig and others is most helpful.[18] This is the Foucauldian conception of power that involves both *constraint* and *enablement*.[19] Actors are said to hold power if they have demonstrated the ability to (1) constrain certain forms of behavior as well as to (2) enable other forms of behavior. This is echoed by Mills who defined the power elite as being "in positions to make decisions having major consequences" and that "whether they do or do not make such decisions is less important than the fact that they do occupy such pivotal positions."[20] It must be stated, however, that also central to our understanding of governance is the importance of *intentionality*. Bertrand Russell is famous for arguing that power is "the production of intended effects,"[21] and considering the level of intentionality of potential governing actors is extremely important for our discussion insofar as intentionality signals causality. We want to be able to distinguish between those actors who are structurally positioned to make decisions and create policies with intentional effects versus those who can be weeded out from the governance discussion because their role in causality is hazy, at best.

If such a definition for power is utilized then identifying who holds power on the Internet can be answered more scientifically. The goal of this research is to identify those actors who simply have influence in the policy process versus those who have repeatedly provided evidence of their decision-making authority through policymaking. Who has *influence* versus who has *authority* is a critical distinction.

With regard to the Internet, it follows that *governance can be defined as the practical exercise of decision-making authority through a demonstrated ability to create policies that constrain or enable behavior with intentional effects*. Recurring throughout the existing Internet governance literature is the idea that governance is the persistent shaping of the environment through explicit decision-making.[22] We will build on this notion to show that the Internet's policies—inclusive of policies not only made by governments, but by various private actors as well—are authoritative insofar as they meet the criteria of our definition above, and that empirical evidence comes in the form of existing statements of policy intent that correlate with evidence of policy actions. Actors who have the decision-making authority to create policies that effectively constrain or enable Internet behavior can reasonably be said to govern.

Determining who has this ability to govern through policymaking can further be analyzed by examining what Marcus Franda has called "single controlling points."[23] We will examine the numerous "single controlling points" on the Internet where behavior is constrained or enabled—examples include the Web hosts that operate servers, the Internet Service Providers (ISPs) who deliver Internet access, the websites that control user accounts through Terms of Service (TOS) agreements, and the local and national governments who can still assert their territorial jurisdiction. By analyzing exactly where Internet policies are being created that intentionally constrain or enable behavior, it is here where our inquiries for determining governance will focus.

This interpretation of governance, then, refers to coordinated efforts among various types of actors operating at multiple levels in their efforts to achieve desired ends. Because of the complexity involved, what we will refer to as the Internet's "political architecture" is a mapping of power and authority that includes the relationships among various institutions and other influential actors and policymakers who are best positioned to directly affect change in their environment. This is why our discussion encompasses the full governance spectrum, and not merely the public policies that are made and enforced by formal governmental institutions. Governments and the public sector are limited in their policymaking capabilities as a result of, first, the global dimension and "borderlessness" of the Internet, second, the decentralized architecture of the environment, and third, the limits of technological capabilities. These, along with a unique developmental history characterized at least as much by grassroots movements as by governmental agencies, are the reasons Internet policymaking is differentiated from, by comparison, other traditional policy venues occurring in real-space.

Defining governance in this way helps to place the title question at the heart of this study in context. For years, legislators of governments around the world have often grown frustrated when trying to transpose their authority in their attempts to regulate Internet content and behavior. Problems inevitably arise involving territorial jurisdiction and frequent anonymity achieved through technical measures, and, as a result, many such governmental policymaking processes and implementation strategies have been rendered largely ineffectual. Attempts by U.S. national, state, or local governments to generate policies using a strictly vertical governmental approach have largely been ineffective at achieving desired ends—thus relegating such policies to the status of being merely symbolic actions. Rather, policies of governance, emphasizing coordination among various public, private, and hybrid institutions at every stage throughout the policy process, have become the primary mechanisms for constraining and enabling different aspects of Internet behavior. To be clear, governments are still extremely relevant and essential in the policy process. However, the role of formal governmental institutions has often been fundamentally transformed in the Internet sphere to that of leading coordination-based strategies, acting as a policy catalyst for private-sector actions, or formalizing and legitimating previously made policy decisions after other actors had already propelled the policymaking process forward.

WHAT DO WE MEAN BY "THE INTERNET"?

The Internet is a rather generic term that often means very different things to different people. So in asking the question, "Who governs the Internet?" we need to clarify exactly what it is we are referring to.

In terms of a functional definition, the Internet is a global decentralized network of computer networks, each of which is independently managed in whichever ways its administrator deems fit. Decisions, particularly over technical protocols, are often made by "rough consensus," and their implementation relies completely on voluntary measures being adopted in order to facilitate reliable interconnection and communication. Moreover, the term refers to both the hardware and software components that connect the various networks and computing devices to each other.

In conceptual terms for our discussion of governance, the various entities and ideas that together form the basis of the Internet must be deconstructed into their constituent parts in order to analyze what specifically is occurring with regard to governing the Internet as a whole.

The model we propose in addressing this problem for explaining governance of the Internet is based on the conceptual scheme first put forth by economist and legal scholar Yochai Benkler.[24] This framework conceptualizes communications systems into three layers: the physical architecture, the logical infrastructure (or the code), and the content layers. Benkler originally devised this scheme to understand structural media regulation, arguing that emerging modern network technologies make a decentralized and democratized information environment possible—"enabling small groups of constituents and individuals to become 'users' (or participants), rather than simply 'passive consumers'." Benkler's three layers were conceived as a means of presenting "a new set of regulatory choices" that governments have in decentralized networked environments, and though pertaining primarily to media regulation, I argue that they are valuable for conceptualizing entire modern information communications systems, including the full reach of the Internet itself.

Benkler's framework was later applied by Lawrence Lessig, who used the three-layer model to argue that the Internet "mixes freedom and control at different layers." In his attempt to assess notions of property rights and "the commons" in cyberspace, Lessig extended Benkler's model in two fundamental ways. He utilized the three layers as a way of conceptualizing the Internet specifically, and he used them as a lens for analyzing systems of control—what is free, what is shared, and what is owned in cyberspace.[25] This is particularly important for our purposes in determining governance.

My proposal is to build upon this framework, yet also modify Benkler and Lessig's code layer to create a new distinction *within* the code layer. This study will demonstrate that, when identifying the various actors and institutions involved in Internet governance, two fundamentally different types of actors emerge within the code layer, and therefore it is important to draw this distinction in order to formulate a better understanding of governance arrangements. This will be done by emphasizing the difference between code, understood as technical protocols, versus code as the software developer's tool for creating applications which the end-user encounters. The result is the emergence of what may ultimately be deemed a fourth layer, separating the code layer of Benkler into a protocols layer and an applications layer. This will highlight not only the differences between institutional actors who either create technical protocols or create private, proprietary web applications, but also the different *types* of actors involved in decision-making.

Thus, in contrast to Benkler's three-layer model of (1) the physical architecture, (2) the logical infrastructure (the code), and (3) the content, I propose a new model be introduced that aims to conceptualize the Internet into four layers:

1. The Infrastructure
2. The Technical Protocols
3. The Software Applications
4. The Content

The purpose of this four-layer model is to create a lens for policymakers who seek to produce intentional effects, and this is accomplished by breaking down the different political dynamics at each layer so that policymakers' goals can be better aligned with implementation strategies. These various political dynamics will then be analyzed by addressing three questions within each layer: (1) Why is it important? (2) Who governs it? (3) How are policies being made within it?

By building upon this framework, the task of determining who governs the Internet becomes far more manageable. Public policies and governing efforts at each individual layer, examined independently and separate from one another, can be more clearly ascertained as coherent strategies and tangible entities. Actors at each layer are readily identifiable, and their roles in the policymaking process provide a greater capacity for reasonable analysis. In other words, my approach to answering, "Who governs the Internet?" will be broken down into "Who governs at each layer?" and "Who is governing across layers?"

LITERATURE REVIEW

The field of Internet governance is relatively new by academic standards having only just emerged in the past two decades, and has been developed by a strikingly multidisciplinary cross-section of scholars originating from the fields of law, economics, public administration, international relations, and more. Books on the subject loosely use the terms "govern," "rule," "regulate," and "control" almost interchangeably, which belies the point that governing is based on a more complex political architecture of authority. Understanding both the technical and political dimensions across disciplines is vital. Too often policymakers draft regulatory laws applying to Internet technologies with little understanding of the technologies themselves. Likewise, far too few programmers of such pervading technologies have any involvement in, or knowledge of, the legal systems or the political systems which they are so deeply affecting. The intention here, therefore, is to help bridge this gap by building upon the existing literature across disciplines to develop a new framework central to understanding the governance on the Internet.

There are two general approaches that scholars have used to study Internet governance and public policy: (1) How the Internet is reshaping government and politics, and (2) How government and politics are reshaping the Internet. Our focus shall be on the latter.

The academic literature exposes several distinct arguments in answering how the Internet is being governed. There is an evolution of ideas in answering the question of who governs the Internet—and the wide range of answers include code, national and local governments, international regimes, self-regulation, private engineering consortium groups, and more. Each of these not only serve as a potential counterargument to what will be presented in the chapters that follow, but they also help frame the scholarly evolution of the debate, placing the discussion to follow in better context.

In his path-breaking scholarship isolating architecture as a constraint on behavior online, Harvard Law Professor Lawrence Lessig famously argued that *code* governs cyberspace, meaning that software is programmed to set the rules for behavior, and therefore code and its designers are the central authority.[26] This "code governs" argument is extremely insightful in emphasizing how, in digital environments, technical decision-making has inherently political consequences. Because code itself is an agent of authoritative power—constraining and enabling behavior by defining what actions are even possible in a given space—programmers have a disproportionate amount of authority at several of the single controlling points already mentioned. For example, whether it is controlling the operations of web servers, setting the TOS on social media websites, or establishing network bandwidth caps, programmers make binding decisions over the private virtual spaces that all users of their service must adhere to. They may not completely have free rein—again, the dynamic is too complex than to say any one group of actors controls everything and can do whatever they choose—but in their ability to write code to shape the environment, these programmers definitely prove themselves to be a large part of the governing equation.

However, the great limitation of this argument is that Lessig—to his credit—only claims that code governs cyberspace; not the Internet as a whole. This is a crucial distinction often misunderstood. Though commonly used as a synonym for the Internet, the term "cyberspace" actually refers to only one aspect within the Internet—the virtual environment where people interact with one another and where content, such as websites, images, ideas, and experiences, proliferate. As scholar David Bell has explained it, cyberspace is a cultural artifact—a "product of and producer of culture simultaneously." It is the part of the Internet that "is lived."[27] By contrast, the Internet itself is a communications network defined by its physical infrastructure comprised of wires and cables connecting devices. Its hardware can be found at specific geographical locations; it can be touched. To briefly put this in context, it can be said that someone may post a digital video to a website in cyberspace so long as their computer remains connected to the Internet. Lessig is correct in asserting that code governs cyberspace because, in this context, that is where code is deployed. However, code and its programmers play a far smaller role in the governing dynamic when examining different aspects of the Internet—namely, for instance, the regulation of the physical infrastructure. This is the great limitation of the "code governs" argument—it is immensely valuable for understanding how policies get made regulating cyberspace, but not comprehensive enough to apply it to the Internet as a whole.

A contrasting argument was put forth by legal scholars Jack Goldsmith and Tim Wu who countered with the proposition that *local and national governments* increasingly govern cyberspace, as such governments have begun taking more proactive roles in formulating vertically designed public policies affecting cyberspatial content.[28] Again the focus is on cyberspace, however their narrative suggests that national and local governments derive their power from an already-existing and clear ability to regulate the physical aspects of the Internet—notably, through a re-assertion of their territorial jurisdiction. By leveraging their authority over the physical world—and, hence, the physical infrastructure of the Internet within their sovereign borders—but applying it to regulating content in cyberspace, Goldsmith and Wu are significantly taking a "cross-layer" approach, albeit in a limited fashion, by seeking to explain how authority over one aspect of the Internet can translate into powerful consequences in another. This is an important point that will be revisited shortly.

Meanwhile, international relations scholar Marcus Franda argues that the Internet is governed by an *international regime* consisting of both the public and private sectors, and formalized through international agreements between governments.[29] This is certainly a more comprehensive view of the Internet in its totality, and it utilizes a similar definition of governance based on coordination among multiple actors at multiple levels. However, Franda explores Internet governance from a strict international relations perspective, and as a result his conclusions focus almost exclusively on formal institutions and organizations at that level. Ultimately, his approach is a comprehensive model that can be applicable to the Internet as a whole, but by under-emphasizing the role of individuals and grassroots efforts that have historically played a vital role in driving the Internet's evolution forward, his argument doesn't portray the full picture of power arrangements and policymaking efforts that are occurring in venues other than at the international level.

Then there is the aforementioned Milton Mueller and the cadre of scholars who define "Internet Governance" in the more narrow terms of its technical functionality, arguing that ICANN, the IETF, and a small handful of semi-public *international consortium groups* comprised mostly of academics and engineers govern the Internet, focusing on the administration of the DNS system and standards-setting processes. This narrow definition certainly lends itself to solving the problem of who creates policies regarding the functional, day-to-day operation of the Internet—and international consortium groups like ICANN clearly demonstrate a decision-making authority in that realm. Unfortunately for our purposes, defining Internet governance in this way is generally unhelpful for understanding any Internet issue area other than those focused on technical functionality.

Countering these different notions of code or some type of public or private institutions governing the Internet are proponents who argue that the Web is increasingly *self-regulated* by the masses of users, or netizens, who actively engage in cyberspatial activities and social networks. There are several problems with this argument. First, again, the argument is only intended to apply to cyberspace; not the Internet as a whole. Second, and more importantly, even just as the argument applies to cyberspace, there is a seemingly endless list of examples that contradict the notion that self-regulation is what is currently taking place. The anarchic vision of cyberspatial behavior having a complete lack of oversight is more a part of Internet mythology than it is reality. When people visit websites, they are subject to several single controlling points such as the rules of the website, the web server, the ISP, the telecommunications carrier, and the government or governments who can claim territorial jurisdiction. As will be demonstrated repeatedly throughout this project, self-regulation is a normative, not empirical, depiction of Internet governance today.

Furthermore, much of the debate identifying who governs the Internet has centered on such normative issues of alternative cyber ideologies regarding systems of control. The libertarian model for Internet governance was famously crystallized in John Perry Barlow's classic *Declaration of the Independence of Cyberspace* in 1996, calling for governments of the world to completely stay out of cyber affairs, and that "self-governance" by users will inevitably arise.[30] However, scholars like Barbrook and Cameron have offered direct challenges to the cyber libertarian model, dissecting the principle components of the "Californian school" by seeking to expose it as little more than "an incursion of capitalist values."[31]

There is a longer literary history concerned with the political nature of technologies. As it relates specifically to an Internet context, this is embodied by the debate over how technological systems institute control and order in people's online activities.[32] The architecture of the Internet enables and constrains certain forms of political behavior, and therefore that technical architecture and the policies which sustain it must be viewed as inherently political as well.[33] This is a major point that ought not to be undervalued. In the context of Internet policymaking, *technical decisions often have very political consequences, and thus are often political actions in and of themselves*. As will be explored throughout this study, decisions over which technical protocols to adopt or what type of software code to create have a direct effect on setting the rules for what types of behaviors are even feasible in different cyber spaces, and such decision-making, therefore, inevitably embodies certain political values at the expense of others.

It is in this vein of the technical becoming political that Lessig's "code is law" argument gains so much credence. He purports that just as laws regulate behavior in real-space, code regulates behavior in cyberspace, as "the software and hardware that make cyberspace what it is *regulate* cyberspace as it is." Technology is powerful but not uncontrollable, Lessig notes; it can be designed by human intervention to embody certain values. In the final analysis, cyberspace is made of code, created by people. How people write that code—the type of architecture they set up to protect certain values—will determine if cyberspace will become "free" in the libertarian sense, or "regulable." Indeed, he claims, the invisible hand of cyberspace, guided by commerce, has already constructed an architecture based on control and highly efficient regulation.[34]

As to some examples of when code is law, Lessig cites (1) how in some places you must enter a password before you gain access, while in others you can gain access whether identified or not; (2) how in some places the transactions you engage in produce traces that link those transactions back to you, while in others this link is achieved only if you want it to be; or (3) how in some places you can encrypt your communications, while in others encryption is not an option.

However, Tim Wu formulated a direct counterargument to Lessig's "code is law" argument, publishing an article in the Virginia Law Review actually titled, "When Code Isn't Law."[35] He asks, if the goal is to understand the net effect of code's regulatory forces, how can we not examine the reaction to those forces? In other words, code only has the effect of law if it is largely being complied with, and in cyberspace compliance is certainly not always a given. Rather, he argues, code is more a mechanism for avoidance of the law than it is for change, or even a form of law itself. As he states, "Nothing the code designer does rewrites laws. Instead, code design defines behavior to avoid legal sanctions." The examples he cites to illustrate how code is actually used for avoidance of the law include (1) virtual child pornography, (2) overseas gambling, (3) junk e-mail, and, (4) P2P file-sharing. Thus, according to Wu, code isn't law because, although it can influence the success or failure of a law's effects, it is more accurately viewed as a tool that interest groups use to avoid legal sanctions or use for legal advantage.

Aside from the debate over to what extent code is, or isn't, a type of law, there is also a challenge to the Mueller-led techno-centric approach that focuses on the DNS system and standards-setting processes. Scholars such as Richard Collins have emphasized how, despite the Internet being a global medium, most of the scholarship takes the United States' experience as its focus. While conceding the value of much of this work, Collins writes that, "the idiosyncrasies of the U.S. has, misleadingly, constructed a world of for-profit domain name registries, fretting about network neutrality and the like as a global experience. It is not." He further goes on to highlight three myths of Internet governance that are commonly made in the academic literature: (1) that Internet governance works best when the market decides; (2) that self-regulation is both pervasive and effective (national policies are only marginally important); and (3) that the Internet regulatory environment is distinct from legacy media.[36]

The main problem with the body of Internet governance literature to date is that each of these approaches ultimately leads to a far too narrow understanding of the governance of the entire Internet. Internet governance, particularly viewed through a policy lens, is far too complex to suggest that there is just one answer to the question—akin to one single individual or conspiracy of organizations behind the magic curtain pulling all the levers.

The aforementioned literature either focuses on only one particular aspect of the Internet or oversimplifies a very complicated topic in order to arrive at a single coherent answer. In the former case, it leaves the reader unsatisfied; in the latter, unconvinced.

THE POLITICAL ARCHITECTURE OF THE INTERNET

There exist different sets of primary actors and political arrangements at each Internet layer. As a result, the policies that govern at each layer often have fundamentally different motivations underlying them and seek to achieve different, and often conflicting, objectives. The consequence of this dynamic has been the emergence of policy processes which often address issues and formulate policy alternatives too narrowly, failing to incorporate all four Internet layers. In my conclusion, I will argue that a more comprehensive policy process involving all of the layers is needed for effective governance of the Internet, and that such a process ought to be open and transparent.

The Internet is, in fact, being governed. Staking out a historical-institutionalist approach, it will be demonstrated that policies have been intentionally developed which have shaped and continue to reshape the Internet itself. Again, direct evidence that governance is indeed taking place can be found in the case of ICANN and the governance of the DNS system.

Not only is the Internet, in fact, being governed, but it is being governed by specific and identifiable networks of policy actors at each of the conceptual layers. Governments and public institutions, private commercial firms, public-private hybrid institutions, international agencies, and various NGOs—including specific interest groups and engineering consortium groups—are all actively involved in coordination-based governance policies.

What this study will seek to accomplish in Part I is an identification of which types of policy actors have decision-making authority—an ability to govern, by our previously stated definition—at each conceptual layer. To be certain, there exist different politics, relevant actors, institutional arrangements, and types of public policies at all four of the Internet's layers. It is their identification that is the primary task at hand.

In Chapter 2, we will develop a brief narrative of the Internet's history from a governance perspective. After reviewing its evolution from being a Defense Department project to being transferred under National Science Foundation (NSF) control to, finally, being largely privatized and commercialized, we will see how all four of our conceptual layers came about chronologically and evolved through very different processes. We will argue that this historical development, including the parallel roles of both the public and private sectors, still has tremendous ramifications for understanding Internet governance in the four conceptual layers today.

In Chapter 3, we will examine governance of the Infrastructure layer of the Internet, consisting of the wires, cables, and airwaves that make up the physical network itself. We will determine that the Internet's wired network is governed by a small handful of private telecommunications firms and cable companies who own and operate the infrastructure, and the national governments around the world that, to varying extents, regulate them, and we will explain the political dynamic using an advocacy coalitions framework. Meanwhile, when it comes to governing the Internet's wireless spectrum, we will assert that the Communications Act of 1934 and the spectrum-allocation auctions of recent years serve to demonstrate how and why the federal government—primarily the F.C.C—is the central governing authority, along with an epistemic community of engineers that is paramount in guiding its decision-making.

In Chapter 4, we will examine governance of the Protocol layer, referring to the technical standards and protocols that facilitate digital communication over the network. We will argue that decision-making authority is held by a small handful of international engineering consortium groups—primarily, the Internet Society (ISOC), its IETF, and the World Wide Web Consortium (W3C)—and we will then analyze the constitutional makeup of these organizations and assert that policymaking is best characterized by the "rough consensus" principle. Finally, we will assert that the decisions over which technical protocols to adopt, and how they are to be designed, are, in themselves, an important form of policy which constrain and enable behavior on the Internet.

In Chapter 5, we will examine governance of the Applications layer, referring to the software applications that enable people to use the Internet. We will illustrate how the code underlying both desktop and web applications is a form of policy itself. These software applications enable and constrain the actions of every Internet user on a technical basis, and thus we will demonstrate how code constitutes a unique type of policy, one in which the environment itself is designed to deny the user even a *capability* to act in defiance. We will then argue that a relatively small handful of the most well-capitalized private commercial software firms govern the Internet's applications the most—and this will be demonstrated based on several usability metrics. Ultimately, we will assert that Lawrence Lessig's "code is law" argument best explains how code constrains and enables Internet behavior, only, we will argue, that the code written by private commercial firms often indicates an implicit recognition of the sovereign authority that traditional governmental institutions retain over them.

Finally, in Chapter 6, we will examine governance of the Internet's Content layer, the most highly visible and controversial layer of them all. By highlighting several prominent issue areas such as the regulation of pornographic material online, efforts to mitigate spam, and the regulation of file-sharing over peer-to-peer (P2P) networks, we will argue that while national governments certainly have governing authority over Internet content to an extent, ISPs and private website operators (through their TOS Agreements) also have demonstrated their authority to make policies that directly constrain or enable behavior with intentional effects, particularly in the transnational context.

Fundamentally, these layers are not sequential, nor are they necessarily mutually exclusive. Policies made at one layer typically have significant consequences for shaping the policy environment at the other layers. For example, at the Protocol layer, the decision to adopt the Transmission Control Protocol/Internet Protocol (TCP/IP) protocol, which is open and universally accessible, rather than alternatives that may have allowed for far more centralized control, directly led to the development of the open and decentralized Internet that currently exists. If decisions had been made to adopt more closed, rather than open, standards and protocols, the policy environment affecting cyberspace at, say, the Content layer would be fundamentally different, allowing for greater government-imposed systems of control—as centralized authority would be built into the technology itself.

As a result, from a prescriptive point of view, I will argue that *policy objectives can be best achieved by either identifying which layer is most appropriate to a specific problem and designing narrowly targeted policies with the context of that specific layer's political dynamics in mind, or by targeting one layer with the direct intent of causing cascading effects at another layer entirely*. In other words, whether policymakers choose to work within the political architecture of one specific layer, or whether they choose to take a cross-layer approach seeking cascading effects, either way it is the conceiving of Internet-based problems in terms of our conceptual layers that will ultimately prove to be a valuable tool for policymakers. Doing so will enable the development of better Internet policies that can more reliably achieve desired outcomes.

Policymakers ought to utilize this conceptual model because it accounts for the Internet's complexities, both in technical and political terms. The four-layer model, and its resulting map of political architecture creates four distinctive policy arenas, each with its own set of criteria for determining what policy designs are most appropriate, and each with its own political dynamics that will ultimately influence to what extent a policy will be effective in achieving the desired outcomes. The question is as old as Political Science itself: If something needs to get done, who has the power to do it? The four-layer model and its resulting map of political architecture provide the answer.

Table 1.1 Summary of the Internet's Political Architecture

LAYERS		*Why is it Important?*	*Who Governs?*	*How Are Policies Being Made?*
	Infrastructure	Enables the actual connection between network devices	National governments, private telecom firms	Wired: Advocacy Coalitions; Wireless: Epistemic Communities
	Protocols	The languages by which devices communicate over the network	International engineering consortium groups	"Rough Consensus" principle
	Applications	The tools which allow people to make use of the network	Private commercial software firms	"Code is law" principle
	Content	The actual material that people see, read, listen to, download, watch, and interact with while online	Private ISPs, hosting companies, website operators, and national and local governments	TOS Agreements, Issue Networks

THE CASE OF U.S. NATIONAL CYBERSECURITY POLICY

In Part I of this study we will explore each of the four conceptual Internet layers—the Infrastructure, the Technical Protocols, the Software Applications, and the Content. At each, it will be ascertained why that layer is important, who governs it, and how are policies being made that affect it. Viewed in its totality, this will define the current political architecture of the Internet.

In Part II we will apply this new four-layer model and resulting political architecture by performing a detailed case study on U.S. national cybersecurity policy, post-9/11. As will be demonstrated, this case is a prime example both of what works and what doesn't when policies are designed to coordinate actions among governments, private commercial firms, hybrid institutions, and the software and engineering communities—in other words, within the context of the political architecture that will be laid out.

The story of U.S. cybersecurity policy can be thought of in two parts. First, in the initial years following the terrorist attacks of September 11th, 2001, the story is about the policymaking process that ultimately led to the *National Strategy to Secure Cyberspace (NSSC)* policy document. Second, in the years since, the story is about the formation of a new bureaucratic regime headed by the U.S. Department of Homeland Security (DHS).

Our objective will be to utilize our four-layer model and its resulting map of political architecture by analyzing the issue of national cybersecurity from a broad public policy perspective in order to test the hypothesis, and commonly held perception, that cybersecurity policy's failures are the result of a flawed policy design that focuses almost exclusively on voluntary public-private partnerships.

First, we will conduct a descriptive analysis of the problem definition underlying the issue. The generalized problem which U.S. national cybersecurity policy is designed to address—namely, digital threats to the nation's critical cyber assets—can be made more specific by deconstructing the problem using a layer-based approach. At the Infrastructure layer, the threats include outright destruction of the Internet's physical components, such as critical telecommunications lines or operating centers, and the hijacking of industrial control systems, such as regional power grids. At the Applications layer, the threat is comprised of malicious code infiltrating vulnerable software applications to steal data or hijack network devices. At the Content layer, the threat comes in the form of defacement of websites or websites being taken offline completely.

The problem definition will be further analyzed by highlighting the categorical and specific mechanisms by which threat agents pursue their goals at each of the aforementioned layers. We will introduce a new typology that draws important distinctions between cyberterrorism, hacktivism, cracktivism, and cyberwarfare, and place specific deployment mechanisms like viruses, worms, botnets, and distributed denial-of-service (DDoS) attacks in this context. Again, our objective is to clarify the problem that cybersecurity policy is designed to address, and conceptualizing this complex, often vague, problem in terms of layers will prove useful in understanding the subsequent policy analysis.

Second, we will perform a detailed analysis of the primary document currently guiding U.S. national cybersecurity policy—the Bush administration's *NSSC*.[37] The policy design of this document is important in how it implicitly addresses all four layers in our conceptual framework. It calls for enhancing the protection of the nation's critical cyber assets by bolstering the defenses of the physical infrastructure, and directly references how this can be achieved through designing more secure technical standards and protocols, promoting more secure software application development in the private commercial sector, and by patrolling Web content.

Third, we will examine the policymaking process that led to the *National Strategy*. This process can be characterized as open, but flawed. A presidential advisory board released 53 questions to the public for comment, then drafted an initial proposal which was discussed in several town hall meetings across the country, ultimately leading to the final version of the policy. It was heavily influenced at every stage by large private corporations, and from the outset of its implementation it came under heavy criticism for failing to allocate enough resources to the problem and for relying on a strictly voluntary public-private approach. Implementation was further hindered by a high turnover rate at the top levels within the newly created Executive bureaucracy—the DHS's National Cyber Security Division (NCSD). As we will demonstrate, this policymaking process was inclusive of most of the major governing actors set forth in our political architecture (and that in itself is significant), however organizational conflicts between them, again contextualized in terms of who has authority at each specific layer, played a large role in derailing the policy's implementation.

Next, we will seek to clarify the current bureaucratic regime governing U.S. national cybersecurity policy. As will be explained, this regime had been headed primarily by the NCSD division within the DHS, however, following a weakened period of having conflicting roles with the newly created National Cyber Security Center (NCSC), the NCSD is now competing intensely to retain its governing authority with the Department of Defense (DoD) and the National Security Agency (NSA), particularly the military's CYBERCOM command center.

Finally, we will then attempt to tie all of this together by examining cybersecurity policy in action—namely, what actually happens in the face of a cyberattack. What becomes evident is the centrality of the private sector, particularly in preventing attacks; also, the reliance on software applications and technical protocols both in prevention and response, particularly network-monitoring tools and specific antivirus products; and finally, that the federal government's role is relegated primarily to being a coordinator among private actors. U.S. Computer Emergency Readiness Team (US-CERT) is vital to raising awareness about cyberattacks and for information-sharing, but ultimately, U.S. national cybersecurity policy thus far limits the federal government from taking more forceful measures beyond that point. The four-layer conceptual model again proves helpful in contextualizing both the problem stream and solution stream surrounding the issue by framing it in these terms.

Ultimately, by applying our four-layer model and its resulting map of political architecture to the issue of U.S. national cybersecurity policy, we will argue that its overriding policy design and policymaking process are reflective of how all four conceptual layers are important in their own right, and that this confirms the utility of the four-layer model in general. The acknowledged failures of U.S. cybersecurity policy have more to do with an implementation process characterized by institutional turmoil within the Executive Branch of the federal government than with a flawed policy design or policymaking process—and, in fact, this only serves to reinforce our argument that government alone does not have adequate governing authority to achieve their desired outcomes. Even the common criticism of the NSSC's policy design relying too heavily on public-private partnerships is not so much a flawed design element as it is a recognition of the Internet's decentralized reality where numerous governing actors have authority at different layers. The lessons of U.S. cybersecurity policy reaffirm that the best way to create meaningful Internet policies that can be effectively implemented lies in creating policies that target the layer most appropriate to specific problems in order to produce intentional cascading effects at, what is often, another layer entirely.

In summary, the main purpose of this book will be three-fold: (1) to develop a new conceptual model that deconstructs the Internet into four policy layers; (2) to use this model in formulating a new political architecture that accurately maps out and depicts authority on the Internet—ultimately determining who governs at each layer; and (3) to use the case of U.S. national cybersecurity policy, post-9/11, in order to evaluate the usefulness of both. If we are to answer, "Who governs the Internet?," we need to know how to frame the question, how to answer it, and whether or not our method of framing and our answers are helpful. That is our goal in the following chapters.

NOTES

1. *Joint Project Agreement Between the U.S. Department of Commerce and the Internet Corporation for Assigned Names and Numbers* (September 29, 2006). Retrieved on February 28, 2013 from http://www.ntia.doc.gov/files/ntia/publications/signedmou290906.pdf.

2. "Program Statistics," *ICANN*. Retrieved on June 7, 2015 from http://newgtlds.icann.org/en/program-status/statistics.

3. Ingrid Lunden, "ICANN Applicants for New TLDs Revealed as Part of 'Reveal Day': The Full List," *TechCrunch* (June 13, 2012). Retrieved on February 28, 2013 from http://techcrunch.com/2012/06/13/icann-applicants-for-new-tlds-revealed-the-full-list/.

4. Milton L. Mueller, *Ruling the Root: Internet Governance and the Taming of Cyberspace* (Cambridge, MA: MIT Press, 2002).

5. Laura DeNardis, *Protocol Politics: The Globalization of Internet Governance* (Cambridge, MA: MIT Press, 2009).

6. B. Guy Peters and John Pierre, "Governance Without Government? Rethinking Public Administration," *Journal of Public Administration Research and Theory* 8.2. (April 1998): 223–243.

7. Robert A. Dahl, *Who Governs? Democracy and Power in an American City* (New Haven, CT: Yale University Press, 1961). See page 3: "if there are great inequalities in the conditions of different citizens, then there must also be great inequalities in the capacities of different citizens to influence the decisions of their various governments."

8. Carolyn J. Hill and Laurence E. Lynn, Jr., "Is Hierarchical Governance in Decline? Evidence from Empirical Research," *Journal of Public Administration Research and Theory* 15.2 (2005): 173–195.

9. H. George Frederickson and Kevin B. Smith, *The Public Administration Theory Primer* (Boulder, CO: Westview Press, 2003).

10. Hill and Lynn, 175.

11. Hill and Lynn, 174.

12. Candice Jones, William S. Hesterly, and Stephen P. Borgatti, "A General Theory of Network Governance: Exchange Conditions and Social Mechanisms," *The Academy of Management Review* 22.4 (October 1997): 911–945.

13. Mark C. Taylor, *The Moment of Complexity* (Chicago, IL: University of Chicago Press, 2001).

14. Richard Collins, *Three Myths of Internet Governance: Making Sense of Networks, Governance, and Regulation* (Chicago, IL: University of Chicago Press, 2009), 59.

15. Collins, 60.

16. Piotr Konieczny, "Adhocratic Governance in the Internet Age: The Case of Wikipedia," *Journal of Information Technology and Politics* 7.4 (October 2010): 263–283.

17. Henry Mintzberg, *Tracking Strategies: Toward a General Theory* (Oxford, England: Oxford University Press, 2007).

18. Lawrence Lessig, *Code and Other Laws of Cyberspace* (New York, NY: Basic Books, 1999). Lessig does not explicitly define "governance," but makes references to how code and "architecture regulates behavior" in cyberspace, and to "constraints on how you behave." See Chapter 7.

19. Michel Foucault, *Power/Knowledge: Selected Interviews and Other Writings, 1972–1977*, ed. Colin Gordon (New York, NY: Pantheon, 1980).

20. C. Wright Mills, *The Power Elite* (New York, NY: Oxford University Press, 1956). See pp. 3–4: "Whether they do or do not make such decisions is less important than the fact that they do occupy such pivotal positions: their failure to act, their failure to make decisions, is itself an act that is often of greater consequence than the decisions they do make. . . . Often they are uncertain about their roles, and even more often they allow their fears and their hopes to affect their assessment of their own power. No matter how great their actual power, they tend to be less acutely aware of it than the resistances of others to its use."

21. Bertrand Russell, *Power: A New Social Analysis* (London, England: Allen and Unwin, 1938).

22. Laura DeNardis offers a helpful definition of Internet governance referring to "policy and technical coordination issues related to the exchange of information over the Internet" in the context of architecting civil liberties into IPv6 protocol design. See Laura DeNardis, *Protocol Politics: The Globalization of Internet Governance* (Cambridge, MA: MIT Press, 2009).

23. Marcus Franda, *Governing the Internet: The Emergence of an International Regime* (Boulder, CO: Lynne Rienner Publishers, 2001).

24. Yochai Benkler, "From Consumers to Users: Shifting the Deeper Structures of Regulation," *Federal Communications Law Journal* 52 (2000): 561–563.

25. Lawrence Lessig, *The Future of Ideas: The Fate of the Commons in a Connected World* (New York, NY: Vintage Books, 2002).

26. Lessig, *Code.*

27. David Bell, *An Introduction to Cybercultures* (New York, NY: Routledge, 2001), 2.

28. Jack Goldsmith and Tim Wu, *Who Controls the Internet? Illusions of a Borderless World* (New York, NY: Oxford University Press, 2006).

29. Franda.

30. John Perry Barlow, *Declaration of the Independence of Cyberspace* (1996). Retrieved on January 5, 2006 from http://www.eff.org/~barlow/library.html.

31. Richard Barbrook and A. Cameron, *The Californian Ideology* (1996). Retrieved on August 23, 2012 from http://www.hrc.wmin.ac.uk/theory-californianideology.html.

32. Langdon Winner, *Autonomous Technology: Technics Out of Control as a Theme in Political Thought* (Cambridge, MA: MIT Press, 1977).

33. Andrew Shapiro, *The Control Revolution: How the Internet is Putting Individuals in Charge and Changing the World We Know* (New York, NY: Public Affairs, 1999).

34. Lessig, *Code*.

35. Tim Wu, "When Code Isn't Law," *Virginia Law Review* 89.4 (June 2003): 679–751.

36. Collins.

37. *National Strategy to Secure Cyberspace* (February 2003). Retrieved on March 5, 2005 from http://www.defense.gov/home/features/2010/0410_cybersec/docs/cyberspace_strategy%5B1%5D.pdf.

Chapter 2

A Brief History of the Internet as Told Through Four Policy Layers

The Internet of today did not arise by accident nor did it emerge overnight, but rather is the product of a half-century long evolutionary process. In order to place our governance discussion in an appropriate context, it is necessary to understand the Internet's historical development. This history is comprised of two competing narratives—the role of public institutions (including government agencies and government-funded university programs) versus the role of decentralized private stakeholders (encompassing both private commercial firms as well as independent, nonaffiliated individuals). The Internet today is not really a single entity, but rather is more accurately described as a collection of millions of privately owned and operated computer networks, independent from one another, and each with its own set of rules prescribed by its own administrators. How did this structure come about, and why is the historical process still relevant in determining governance today?

Histories of the Internet abound, and it is not our intention to rehash what previous scholars have written about at length.[1] Our approach for this project will be simply to highlight the major events in the Internet's history that directly relate to our governance discussion focusing on the four conceptual layers—specifically, the constant interplay between public institutions and decentralized private stakeholders that has characterized the Internet throughout its history. Rather than simply choosing which lens—public or private—is "correct" and which to disregard (or minimize in importance), this narrative will describe the vital role that both have played in shaping the Internet since its inception. I refer to this approach as incorporating "parallel" narratives because, as will be discussed, although the federal government played a central role in creating and funding the original network, beginning in the 1970s there were essentially multiple networks—some of the nonprofit government-sponsored variety, while others decentralized and privatized—operating simultaneously and in parallel to one another.

It will also be illustrated how the four conceptual layers came about chronologically and evolved through different processes—some, like the Infrastructure, were built upon preexisting legacy systems, while others, like the Protocols, had to be newly created from scratch. As a result, the constant interplay that will be described between the various public and private actors, which continues to the present day, will be valuable in explaining current governance arrangements at each layer, as well as the unique challenges that the Internet continues to pose to policymakers.

Ultimately, this chapter will utilize the historical narrative put forth to develop two key arguments that will recur throughout this project; first, that each Internet layer has unique developmental characteristics including different policymaking processes that have carried over to the present day. Thus, the layers are distinct and they matter. Second, that history has demonstrated how sometimes governments and public policies have had governing authority, while other times the engineering community's technical decisions or other private-sector policies have been authoritative; the central point being that the determination of governance all depends on which layer is being analyzed.

ARPANET & THE INTERNET'S EARLY YEARS

The creation of the Internet can be attributed to a direct response on behalf of the U.S. Government to the Soviet Union's launch of Sputnik on October 4, 1957. As a consequence of Sputnik's launch, the Defense Department issued Directive 5105.15 establishing the Advanced Research Projects Agency (ARPA) in 1958.[2] ARPA's mission was:

> to assure that the U.S. maintains a lead in applying state-of-the-art technology for military capabilities and to prevent technological surprise from her adversaries.[3]

Both ARPA's objectives and funding were supplied by the military, which later renamed the agency the Defense Advanced Research Projects Agency (DARPA) in 1972. Its broad mandate led to a research agenda in pursuit of military goals which were not necessarily limited to military applications. It was at ARPA in the 1960s and 1970s that scientists propelled forward major technological advances in the fields of microelectronics, computing, and network communications.[4] Consequently, the DoD took the primary role in governing the Internet during these early years.

The Internet itself was directly born as a result of the military's desire to be able to analyze a battlefield situation remotely and as the battle was progressing. Joseph Licklider and ARPA's Information Processing Techniques Office (IPTO) had been charged with the task of applying a computer's analytical power to ever-changing battlefield situations, and the problem emerged of how to communicate the results back-and-forth between the command center and the battlefield. Some type of remote access to the computer running the analysis was necessary. In the late 1960s, this led to invention-by-necessity—the creation of the world's first computer network called ARPANET. An interconnection of ARPA's host computers was distributed mostly to universities located across the United States and used existing telephone lines to transfer data. This was additionally significant because, by deciding to utilize the existing telephone network for data communication, the Internet's physical infrastructure was already in place.

On November 29, 1969, the first two nodes of the ARPANET—the University of California at Los Angeles (UCLA) and the Stanford Research Institute (SRI)—exchanged their first message. This was made possible by IPTO's Larry Roberts who made the decision over ARPANET's technical design, choosing among protocols for fully interconnected point-to-point leased lines, line-switched (dial-up) service, or packet switching.[5] The decision was ultimately for a packet-switched system, and the initial host-to-host software used to connect ARPANET's first few host computers would later be replaced by the NCP protocol in 1970, and then eventually replaced with the TCP/IP protocol in 1983.[6] This first transmission was also the result of a contract that ARPA awarded to the private firm BBN to build an interface message processor (IMP)—essentially, a first-generation packet-switched router—outbidding other firms like Raytheon and Jacobie Systems.[7] Within a month of UCLA and SRI's first packet-switched message being transmitted, two additional nodes were added—the University of California at Santa Barbara and the University of Utah. Within a year, the ARPANET was growing at a rate of one new node per month, and in less than two years the planned 15-node network was in place and operational.[8]

Already evident at this early stage are a few noteworthy characteristics that are of great significance to our Internet governance discussion. First, in terms of simply creating a functional ARPANET with a mere 15 nodes, a physical infrastructure of wired telephone lines was prerequisite, and a common language or communications protocol needed to be used by all nodes desiring to function on the network. Thus, our first two Internet layers—the Infrastructure and the Protocols—have already emerged.

Second, the way these two layers have emerged, even at this early stage in the narrative, is instructive for understanding how those layers are governed to the present day. At the Infrastructure layer, the federal government acted as a central coordinator in its capacity of regulating the existing telephone network (this will be explored more thoroughly in Chapter 3). At the Protocol layer, the decision-making process that led to the adoption of a packet-switched protocol—a process characterized by public Request for Comments (RFCs) and seeking a rough consensus among the engineering community—will be the pattern of decision-making used for protocol adoption to the present day (explored further in Chapter 4).

Third, the federal government, private commercial firms, and universities have all emerged even at this embryonic stage as being vital to the ARPANET. In fact, while the 4 original nodes of the ARPANET in 1969 were all located at universities, an examination of the network's next generation of nodes reveal that an institutional mix of universities, government agencies, and private commercial firms were already connected. The ARPANET nodes following the 4 original nodes were MIT, Harvard, BBN Technologies, Systems Development Corp., Stanford, MIT's Lincoln Labs, Carnegie Mellon University, Case-Western Reserve University, NASA/Ames, RAND, and the University of Illinois-Urbana.[9] This reveals ARPANET's public-private hybrid makeup even in 1971, and it is a crucial point that should not be understated.

In order to overcome the skepticism of the people who were being requested to attach their equipment to and use the ARPANET, Larry Roberts made connection to the ARPANET mandatory for all computer centers funded through the IPTO. This policy was motivated largely by cost. By the mid-1960s, ARPA was the major funding source for most of its contractors, and buying equipment for them represented a large expense for the agency. Requiring the various computing centers to connect to the ARPANET would not only pool hardware, software, and data resources more efficiently from a technical perspective, it would also eliminate the need, and corresponding cost, of "wastefully duplicating" the same state-of-the-art resources at each facility.[10] In other words, the ARPANET was a means for sharing resources and avoiding redundancy, and making connection to it mandatory was a way of cutting expenses for the ARPANET's chief benefactor. This is a clear case of the federal government using its procurement powers as leverage to drive the Internet's early growth.

Funding for the ARPANET project was by no means a priority within DoD. As former ARPA Director Stephen J. Lukasik describes, the Agency's previous work had been carefully watched by the Secretary of Defense, the White House, and the president's Science Advisory Committee, but with the ARPANET project Licklider was largely "left alone." Licklider recalled:

> [The Director] seemed too busy, he was just relieved to get somebody to run the office . . . I talked with him periodically [and] he would make suggestions about directions of things, but pretty much let me do what I wanted to do.[11]

However, despite this seemingly high level of independence in guiding the direction of ARPA's research, the politics of funding the Agency within Congress was another matter. Lukasik explains that by 1968:

> The environment for ARPA was quite different. While the need for survivable networks had not changed, political forces were tightening the screws on ARPA. The Agency's budget was shrinking, partly to help pay for the Vietnam War and partly because opponents to the war in the Congress were using every opportunity to cut programs in DoD. "Fraud, waste, and abuse" was a frequently invoked litany of sins. Lack of relevance to national security was another. Senator Proxmire periodically identified projects for a Golden Fleece Award, projects with scientifically accurate but easy-to-ridicule titles. Senator Mansfield was soon to demand "relevance statements" for every R&D line item in DoD.[12]

Nevertheless, funding for ARPA and the ARPANET was ultimately appropriated as part of the larger DoD budget. But it was a result of this sometimes-hostile political environment in Congress that the decisions were made to first implement the small four-node network before expanding to one more robust, and to the very selection of which four universities would be included in it.[13]

The next major event in the Internet's historical development came at the International Conference on Computer Communication (ICCC) in 1972 when ARPA engineers along with associated university faculty and students successfully demonstrated the potential of the network through a number of simulations. The new conference "marked a turning point." Organized by ARPA's Larry Roberts and Bob Kahn with the express purpose of promoting the ARPANET,[14] and attended by approximately 800 computer communication professionals, government employees and academics, simulations demonstrated interactive chess games and "conversations" between computers located at MIT and Stanford, and were accessed from the conference terminals in Washington D.C. Suddenly, the experimental ARPANET, a testbed for networking theory and technology, had been transformed into a functional tool with exciting and practical applications. The ICCC was "for packet switching what the Centennial Exposition in Philadelphia in 1876 was for the telephone: the public unveiling of a technological discontinuity." According to first-hand accounts, people weren't leaving the room until well after midnight every night.[15]

In the month following the conference, ARPANET traffic increased 67 percent.[16] At this point, 30 institutions were connected to the network ranging from industrial for-profit commercial installations and private consulting firms like BBN, Xerox PARC, and the MITRE Corporation, to government sites like NASA's Ames Research Laboratories, the National Bureau of Standards, and Air Force research facilities, to numerous universities.[17]

However, the successful demonstrations at the ICCC conference had significant unintended consequences. Rather than a large movement toward the ARPANET itself, the more general concept and application of computer networking was the real revelation to those in attendance. Following the conference, "people started to apply their newfound appreciation for networking in highly original and unanticipated ways that suited their own needs." While the federal government remained the principal funding source and governing authority for the ARPANET, the newly arising innovative spirit coming out of the ICCC led to a highly decentralized approach to the next developmental stage of networking technology. The direct consequence was a surge in growth of private, local area networks (LANs) that existed outside the ARPANET.

NSF CONTROL AND THE RISE OF PARALLEL PUBLIC AND PRIVATE NETWORKS

Throughout the 1970s, the ARPANET would continue its slow growth under the Defense Department's control, but the spread of thousands of private LANs would occur independent from the government in the private sector. The IPTO could create mandates and set policies for all computer resources linked to the ARPANET due to their procurement power, such as banning the use of the network for commercial purposes, however the IPTO had no such policymaking capability over computer resources connected to networks which were not the ARPANET, and the proliferation of private LANs demonstrated this point as private network administrators were independently making decisions affecting their own networks. At this stage, it is clearly evident that two parallel narratives were both in play—the ARPANET was developing under the direct guidance and funding of the federal government while, simultaneously, thousands of private local networks were developing in a highly decentralized manner apart from it.

A shift in control of the ARPANET would occur in the 1980s. As more universities, research centers, and private entities increasingly used the network for various purposes, the military created a separate network which it named the MILNET where it could, once again, pursue strictly military objectives. As the military migrated to MILNET, control over, and funding of, the ARPANET was transferred from the Defense Department to the NSF.

Meanwhile, more earnest efforts were being made to interconnect the ARPANET with ever-more government agencies and universities as well as with the commercial and more freely accessible private LAN networks that existed in the private sector. A core technical problem arose: the methods and protocols that each LAN implemented were often different from one another, and that was an obstacle to interconnecting all of the various networks together. The solution was the deployment of the TCP/IP protocol in 1983.[18] TCP/IP overcame such differences among the networks by shifting the responsibility for technical reliability away from the network itself and instead toward the host devices.[19] As a result, both the ARPANET and private LANs were able to much more easily interconnect with outside networks. There no longer needed to be a uniform set of technical policies among all networks in order to maintain operability. An Internet of heterogeneous networks became possible.

The great insight to be drawn from TCP/IP's development and ultimate deployment is that, as the demand had grown for existing public and private networks to interconnect, there was no demonstrable governance over all of the networks involved, and thus no central authority to turn to. It became clearly evident that the Internet did not represent the master plan for a global network of any individual, university, corporation, or government. As the case of TCP/IP illustrates, no single body governs the Internet's functional operation, determines what services it will or will not provide, or decides who can or cannot use its resources. The Internet is not one entity; it is a collection of millions of privately owned and operated networks, located across the globe, each setting its own policies for the behavior that will take place on it.

Interconnection of various networks, owned and operated by different entities, meant that there needed to be some agreement over which technical protocols would be implemented in order for the computers on different networks to communicate with one another. As each private network chose which protocols it deemed were best for itself, the drive toward interconnection meant that others would have to follow suit and adopt the same protocols. Some decision had to be made in order to ensure interoperability.

Here is the prime example of how the historical development of the Internet was not driven solely by the military, government, universities, a commercial firm, or any other single entity, or even by a trust-like form of collusion. Decisions were not made by a central authority, but rather resulted from a highly decentralized process involving numerous actors of various types, all seeking to further their own self-interest—in this case, defined as interoperability. As more and more networks adopted certain protocols, particularly TCP/IP, a rough consensus developed over which were to be used on a mass scale. This process, or "rough consensus principle," is still a defining characteristic of Internet governance.[20] No single entity had the authority or decision-making capability to force the adoption of certain protocols, but when a rough consensus by Internet users had developed, if someone didn't adopt those protocols then they simply would not be able to interconnect.

Rather than government institutions or relatively large, well-capitalized private companies rising to the forefront of protocol decision-making, it is during these years that epistemic communities of engineers, embodied in consortia groups like the ISOC and the IETF, came to prevail and, over time, establish their governing authority.[21]

By the late 1980s, the NSF had taken over funding for the operation and continued development of the Internet—funding appropriated by the U.S. Congress and approved by the president—while simultaneous research and developmental efforts were being pursued at universities and at numerous private commercial firms, independent from the NSF. The Defense Department—DARPA, specifically—had voluntarily ceded its governing authority to the NSF as it refocused on its core mission to act as a research agency, not a communications operator. Meanwhile, the NSF built its own network in 1986 to link its supercomputers called the NSFNET, which was different from the ARPANET in one fundamental way. The ARPANET was still a single, homogenous network, whereas the NSFNET consisted of two distinct types of networks—local, geographically regional networks in addition to a central high-speed backbone.[22]

This difference in architectural design would become significant in ushering in ARPANET's demise—not as the result of an intentional government decision to do so, but in response to technological forces and a practical need to maintain functionality and interoperability. Unlike the ARPANET and the rest of the Internet at that time, the NSFNET did not immediately use the TCP/IP protocol, and this was viewed as a large deficiency. In order to make the transition to TCP/IP, its backbone had to be taken offline, and thus, while the change was being made, an arrangement was setup whereby the NSFNET would use the ARPANET as its backbone. This connection, "seemingly insignificant at the time," permanently changed the Internet itself by making the NSFNET available to most universities for the first time. The Internet suddenly included regional network operators as well as a variety of new supercomputer facilities.[23]

An awareness quickly developed that the new NSFNET backbone would be a faster and more ideal solution for both transitioning the Internet to the latest technologies as well as for removing any vestiges of the military's involvement from what had become a largely civilian enterprise. The move to the NSFNET backbone was made with the vast majority of Internet users never even being made aware of the switch, and in February 1990, the ARPANET was permanently shut down because of the rapid exodus that had already begun toward the NSFNET's faster backbone.[24] As with the adoption of TCP/IP, the shutting down of ARPANET illustrates how decision-making was being heavily influenced by technological efficiency and, in this case, faster bandwidth speeds.

PRIVATIZATION, COMMERCIALIZATION, AND THE WORLD WIDE WEB

With the infrastructure and technical protocols now largely in place, the next stage in this historical evolution was ushered in with the privatization and commercialization of the Internet, which came shortly thereafter. Even as control of the Internet passed from the Defense Department to the civilian authority of the NSF, commercial traffic nevertheless remained prohibited because the NSF was still a government agency and its Acceptable Use Policy (AUP) clearly stated that all commercial traffic by for-profit institutions across its backbone was prohibited.[25] This AUP ban on commercial traffic hindered the Internet's growth and its adoption by new classes of users.[26] However, in keeping with an already-established tradition of decentralized development, independent private commercial networks began to emerge to meet the needs that the NSF was not satisfying. As the operations of these independent commercial networks grew, and as some merged to form larger networks, and as others made agreements to share and route traffic between their various networks, the result was an internetwork of commercial TCP/IP networks that paralleled the noncommercial Internet run by the NSF. Once again, the competing narratives of an institutional nonprofit Internet versus a decentralized commercial Internet could be seen simultaneously running parallel to one another. While it is true that the original networking technology of the ARPANET had been sponsored by government funding, the further development of that technology, by the 1980s, had veered off in a number of directions largely due to private capital, and through processes in which the government played no role. These narratives of an institutional nonprofit Internet versus a decentralized commercial Internet are "parallel" because, by this time, both internets were operating simultaneously.

Observing this development, by the late 1980s, the Office of Science and Technology Policy (OSTP)—a presidential advisory agency with the Executive Branch, created by Congress—established committees to propose a process whereby the NSF could transition the Internet from a government-funded operation into a private commercial service.[27] Consequently, in November 1991, the NSF's new Project Development Plan called for Internet service to be delivered by independent and private commercial and nonprofit ISPs, each of whom would operate its own network and backbone.[28]

Ultimately, the committees pushed for full privatization and wholesale commercial use of the Internet amid "little or no public debate."[29] Congressional hearings took place before the Subcommittee on Science of the Committee on Science, Space, and Technology in the U.S. House of Representatives, however these hearings didn't focus at all on the issue of whether or not to privatize, but rather only on the best way to privatize.[30] By the early

1990s, the political climate was one in which telecommunications policy for both political parties was based upon notions of deregulation and competition.[31] Indeed, privatization came to be seen as virtually inevitable as politicians and telecommunications executives alike made it clear that the private sector would own and operate the Internet.[32] Senator Al Gore infamously co-authored the High-Computing Performance Act of 1991 that formalized this privatization and opened the Internet for commercial use by establishing the National Information Infrastructure (NII), later nicknamed the "information superhighway."[33]

Without much public debate or controversy over the decision to privatize the Internet, an important question beckons: Did the federal government simply give away an asset that would soon foster billions of dollars' worth of transactions on a daily basis for free? The answer is yes, so the real question that academics have been preoccupied with is why it wanted to. Scholars such as Shah and Kesan have argued that their reasons for privatization were threefold: first, privatization was the dominant policy approach by both political parties in the early 1990s; second, the NSF was encouraging the regional networks to find commercial customers believing that the revenue from new customers would allow the networks to expand and use the economies of scale to lower costs for everyone; and third, there was a growing expression in the private sector about their desire to send nongovernmental and commercial traffic across the Internet, and for the telecommunications companies to start selling connectivity and infrastructure.[34]

If there was any controversy at all, it was the debate over whether the NSF's policies provided a level playing field for network service providers—certain stakeholders like Performance Systems International (PSI), AlterNet, the Commercial Internet Exchange Association (CIX), and the Electronic Frontier Foundation (EFF) were concerned that a small handful of companies like ANS, IBM, and MCI would be the recipients of a structural competitive advantage if the NSF's network management plan went into effect in its original form. The NSF Inspector General reviewed the management plan[35] and, in addressing those critiques, his revised report led to the Scientific and Advanced-Technology Act of 1992 which gave the green light to private and commercial use of the network so long as it would increase the network's utility for education and research.[36] This privatization policy replaced "the need for the NSFNET backbone, any involvement from the NSF, and, in the process, eliminated entirely the commercial restrictions imposed by the NSF's AUP."[37]

In May 1993, the NSF released a formal solicitation to accommodate and promote the privatization and commercialization of the Internet.[38] This document mandated the creation of four Network Access Points (NAPs), which were ultimately sold via closed bid to Sprint, Pacific Bell, Ameritech, and

MFS. Regional networks would no longer connect to the NSFNET backbone, but rather to commercial providers, which would interconnect via the NAPs. Once this migration was complete, the NSFNET was officially retired in April 1995, and shortly thereafter, the NSF ended its sponsorship of the four public NAPs. By this point, the government had transitioned from contracting out Internet backbone services to allowing the market to fully provide them.[39] The privatization process behind the management of the network was now complete.

Once the Internet was privatized and commercialized, the arrival of the World Wide Web became the "focusing event" which would finally launch the Internet into the mainstream culture.[40] One of the largest complaints about the Internet of the late 1980s and early 1990s was its lack of locatable resources. While it was possible to remotely connect one machine to another over the infrastructure and share resources via the TCP/IP protocol, the user still had to know exactly where to find a particular machine, and then would have to search that machine to see what content was available. Even when the sought-after material was found, customized software often had to first be installed in order to view it. A better type of software application was needed to make the Internet more useful.

Addressing this problem, Tim Berners-Lee, while working at the European Organization for Nuclear Research known as CERN[41] in Geneva, Switzerland, invented the World Wide Web in 1991 in order to allow researchers and other individuals to make their work more readily available—either within a LAN, or across the Internet.[42] It was a killer-app designed as a simple document sharing and publication tool, making use of unidirectional "hyperlinks" as electronic cross-references to interconnect documents. The creation of globally unique identifiers for resources on the Web, known as Uniform Resource Identifiers (URIs), made the search and discovery of such resources far easier. Also, the Web was designed to be accessible to anyone who had web browser software installed and Internet access, regardless of what hardware platform or operating system was being used on behalf of the user.

Although the Internet and the Web have often become synonymous in popular vernacular, they are, in fact, separate entities—and their distinction is particularly noteworthy from a historical development perspective. While the Internet made it possible for computers to communicate remotely with one another, it was the World Wide Web—conceived of as an application that runs on the Internet—that allowed documents to be shared, linked to, and found in a practical manner.[43] The Web proved to be such a killer-app that, with the introduction of Mosaic browser software in 1993, traffic on the Web proliferated at an astounding annual growth rate of 341,634 percent.[44] Furthermore, on April 30, 1993, CERN announced that the Web would be released into the public domain, making it free for everyone, with no fees due—producing a rapid shift away from pay-to-use services that made use of Web alternatives like Gopher.[45]

Beginning, then, in the mid-1990s—using primarily the telephone network *infrastructure*, the TCP/IP *protocol*, and the *application* known as the World Wide Web—what resulted was an explosion of *content*. As of July 2008, according to Google Search, over one trillion unique URLs had been discovered.[46]

Thus, the Internet is truly not an entity that can be rightfully claimed to be owned, operated, or governed by any single entity. It is a decentralized network of networks, each of which is independently managed in whichever ways its administrator deems fit. Decisions, particularly over technical protocols, are often made by "rough consensus," and their implementation relies completely on voluntary measures being adopted in order to facilitate reliable interconnection and communication. Historically, both public institutions and decentralized private stakeholders have played vital roles in guiding the Internet's evolution, and it would be an egregious oversimplification to downplay the role of either. However, as it currently stands today, despite governing authority originally resting in the hands of, first, the Defense Department and, subsequently, the NSF, the contemporary Internet is more characteristically described as being a private-sector phenomenon.

As will be shown in the following chapters, formal governmental institutions continue to play a significant role in setting the rules for Internet behavior, especially regarding the physical infrastructure of the network, but it is private commercial firms and nonaffiliated individual computer programmers who are currently the agents primarily responsible for guiding its numerous innovations. The formal handing-over of governing authority by the NSF in the 1990s, coupled with the advent of the World Wide Web, has led to the current state of the Internet: a fully privatized and largely commercial assortment of computer networks, independent from one another, but with certain shared common interests. This constant interplay between public institutions and decentralized private stakeholders, evident throughout the Internet's entire history, continues to shape the governance discussion today—as will be seen in the next section.

NOTES

1. See Abbate, 1999; Hauben and Hauben, 1997; Moschovitis, Poole, Schuyler, and Senft, 1999; Okin, 2005; Rosenzweig 1998; Salus, 1995; Segaller, 1998.

2. *Department of Defense Directive; Number 5105.15* (February 7, 1958). Retrieved on September 4, 2012 from http://semanticvoid.com/docs/darpa_directive.pdf.

3. "DARPA Over the Years," *U.S. Department of Defense Website* (2002). Retrieved on June 25, 2009 from http://www.darpa.mil/body/overtheyears.html.

4. Janet Abbate, "From ARPANET to Internet: A History of ARPA-Sponsored Computer Networks, 1966–1988" (1995). *Dissertations available from ProQuest.* Paper AAI9503730. Retrieved on August 13, 2013 from http://repository.upenn.edu/dissertations/AAI9503730.

5. Lawrence G. Roberts and Barry D. Wessler, *Computer Network Development to Achieve Resource Sharing, in Proceedings of AFIPS* (AFIPS Press, 1970). Retrieved on September 4, 2012 from http://www.packet.cc/files/comp-net-dev.html.

6. "Birth of ARPANET: 1967–1969," *Cybertelecom: Federal Internet Law and Policy*. Retrieved on September 4, 2012 from http://www.cybertelecom.org/notes/internet_history60s.htm.

7. *Cybertelecom: Federal Internet Law and Policy*.

8. Judy E. O'Neill, "The Role of ARPA in the Development of the ARPANET, 1961–1972," *Annals of the History of Computing, IEEE* 17.4 (1995): 76–81.

9. Michael Hauben and Ronda Hauben, *Netizens: On the History and Impact of Usenet and the Internet* (Los Alamitos, CA: Wiley-IEEE Computer Society, 1997). Retrieved on September 21, 2012 from http://www.columbia.edu/~hauben/book-pdf/CHAPTER8.pdf.

10. Janet Abbate, "Building the ARPANET: Challenges and Strategies," *Inventing the Internet* (Cambridge, MA: MIT Press, 1999): 44–46.

11. Stephen J. Lukasik, "Why the ARPANET Was Built," *IEEE Annals of the History of Computing* 33.3 (2011): 8. Retrieved on September 21, 2012 from http://www.cistp.gatech.edu/publications/files/ARPANETv8.pdf.

12. Lukasik, 15.

13. Lukasik, 15–16.

14. According to a conversation with Bob Kahn three months prior to the conference, ARPA hadn't yet set up any budget for the conference "except what their contractors are essentially willing to give as part of their ongoing research." See Richard W. Watson, *RFC 372: Notes on a Conversation with Bob Kahn on the ICCC* (IETF: Network Working Group, July 12, 1972). Retrieved on September 21, 2012 from http://tools.ietf.org/html/rfc372.

15. James Pelkey, "Entrepreneurial Capitalism and Innovation: A History of Computer Communications 1968–1988" (2007). Retrieved on September 15, 2012 from http://www.historyofcomputercommunications.info/Book/4/4.12-ICCC%20Demonstration71–72.html.

16. J. R. Okin, *The Internet Revolution: The Not-for-Dummies Guide to the History, Technology, and Use of the Internet* (Winter Harbor, ME: Ironbound Press, 2005), 73.

17. "Exhibits: 1973," *Computer History Museum.* Retrieved on September 15, 2012 from http://www.computerhistory.org/internet_history/internet_history_70s.html.

18. Vinton Cerf, Yogen Dalal, and Carl Sunshine, *RFC 675: Specification of Internet Transmission Control Program* (IETF: Network Working Group, December 1974).

19. Barry M. Leiner, Vinton Cerf, David D. Clark, Robert E. Kahn, Leonard Kleinrock, Daniel C. Lynch, Jon Postel, Larry G. Roberts, and Stephen Wolff, *A Brief History of the Internet* (2003). Retrieved on September 15, 2012 from http://www.internetsociety.org/internet/internet-51/history-internet/brief-history-internet.

20. The "rough consensus principle" will be explored in greater detail in Chapter 4.

21. The governing authority of these consortium groups will also be thoroughly explored in Chapter 4.

22. Thomas Narten, "Internet Routing," *ACM SIGCOMM Computer Communication Review* 19:4 (ACM, 1989).

23. Abbate, *Inventing the Internet*, 192.

24. Thomas Greene, Larry James Landweber, and George Strawn, *A Brief History of NSF and the Internet* (National Science Foundation, 2003). Retrieved on September 15, 2012 from http://www.nsf.gov/od/lpa/news/03/fsnsf_internet.htm.

25. "NSFNET Acceptable Use Policy," *National Science Foundation Annual Report 1988* (Washington, D.C.: U.S. Government Printing Office, 1988). Retrieved on September 21, 2012 from http://old.cni.org/docs/infopols/NSF.html.

26. Brett A. Perlman, "Pricing the Internet: How to Pay the Toll for the Electronic SuperHighway," *CSIA Discussion Paper* 95–01 (Kennedy School of Government, Harvard University, March 1995).

27. *The Federal High Performance Computing Program* (Office of Science and Technology Policy, September 8, 1989). Retrieved on October 4, 2008 from http://www.ostp.gov/cs/about_ostp.

28. Abbate, *Inventing the Internet*, 37–38.

29. Okin, 105.

30. "Management of NSFNET," *Hearing before the Subcommittee on Science of the Committee on Science, Space, and Technology, U.S. House of Representatives* (102nd Congress, Second Session, March 12, 1992).

31. Dick W. Olufs III, *The Making of Telecommunications Policy* (Boulder, CO: Lynne Rienner Publishers, 1999).

32. Rajiv C. Shah and Jay P. Kesan, "The Privatization of the Internet's Backbone Network," *Journal of Broadcasting and Electronic Media* 51.1 (2007): 93–109.

33. *Public Law No:* 102–194*: High-Performance Computing Act of 1991*. Retrieved on September 19, 2012 from http://thomas.loc.gov/cgi-bin/query/z?c102:S.272.ENR.

34. Shah and Kesan.

35. *Review of NSFNET* (Office of the Inspector General, National Science Foundation, March 23, 1993).

36. *Public Law No:* 102–476*, 43 U.S.C. 1862(g): Scientific and Advanced-Technology Act of 1992*. Retrieved on April 24, 2011 from http://thomas.loc.gov/cgi-bin/bdquery/z?d102:S.1146. The bill was co-sponsored by Senators Mikulski (D-MD), Cochran (R-MS), Sanford (D-NC), & Sarbanes (D-MD), and proceeded through the Senate Labor and Human Resources Committee by Unanimous Consent, passed Without Objection in the House, and was signed by President George H. W. Bush.

37. Okin, 105.

38. *NSF 93–52: Network Access Point Manager, Routing Arbiter, Regional Network Providers, and Very High Speed Backbone Network Services Provider for NSFNET and the NREN(SM) Program* (National Science Foundation, May 6, 1993). Retrieved on September 19, 2012 from http://www.nsf.gov/pubs/stis1993/nsf9352/nsf9352.txt.

39. Shah and Kesan.

40. Scholars have long suggested that focusing events have the potential to arouse a normally indifferent public and set policy agendas. See John W. Kingdon, *Agendas, Alternatives, and Public Policies*, 2nd ed. (New York, NY: Longman, 1995).

41. In French: *Organisation Européenne pour la Recherche Nucléaire.*

42. Tim Berners-Lee and Robert Cailliau, *WorldWideWeb: Proposal for a HyperText Project* November 12, 1990. Retrieved on August 12, 2013 from http://www.w3c.org/Proposal.html.

43. In terms of our conceptual layers, the Internet refers to the Infrastructure and the numerous Protocols that can be used over it; a "network of networks." In contrast, the Web is just one of the ways that information can be sent over the Internet, and relates more to the Applications and Content that are sent via HTTP.

44. Okin, 110.

45. *Ten Years Public Domain for the Original Web Software* (CERN Web Communications—ETT Division, April 30, 1993). Retrieved on September 19, 2012 from http://tenyears-www.web.cern.ch/tenyears-www/Welcome.html.

46. Jesse Alpert and Nissan Hajaj, "We Knew the Web Was Big. . .," *The Official Google Blog* (July 25, 2008). Retrieved on March 19, 2011 from http://googleblog.blogspot.com/2008/07/we-knew-web-was-big.html.

Part II

THE FOUR-LAYER CONCEPTUAL MODEL & THE POLITICAL ARCHITECTURE OF THE INTERNET

In order to answer, "Who governs the Internet?" we must first clarify what exactly it is that we're referring to. As the historical narrative previously explained, the Internet is not a single entity, but rather is a collection of many privately owned and operated networks, each with its own set of policies and decision-making processes. As such, if we are to reasonably go about trying to answer, "Who governs the Internet?" we must break down the question into its constituent parts. Who governs each different aspect of the Internet?

This can best be answered by introducing the concept of layers. It is a term used throughout the academic literature on the subject, and each layer is meant to refer to a different aspect of the Internet—typically either a physical or logical component. Because our focus for this project is adopting a policy-based approach, not only will we define layers substantively in terms of their physical and logical dimensions, but we will also conceive of them in terms of distinct issue areas for policymaking. Deconstructing the Internet into conceptual layers in this manner will help us identify distinct issue areas which ought not to be lumped together in discussions of policymaking, for each layer has its own distinct set of problems, solutions, and governance arrangements. Only when taken together do these layers form a complete vision of the Internet and all of the activity that takes place on it.

The following chapters will examine each conceptual layer of the Internet using a policymaking approach. By analyzing numerous case studies and using both public policies and private-sector Internet policies as its primary units of analysis, it will be demonstrated which actors hold primary governing authority at each layer—that is, who has proven the ability to create policies that constrain or enable Internet behavior with intentional effects—and, additionally, what types of policymaking processes are evident at each layer. Ultimately, this study will argue that the resulting map of the Internet's political architecture is an accurate depiction of Internet governance today. It is something that can be meaningfully utilized not only for better understanding the core typological differences among Internet issues, but also the consequences of those differences. Finally, the arguments will be made such that effective Internet policies are those which target the layer most appropriate to a particular problem, and also that targeting one particular layer as the object of policymaking and/or policy implementation is often an effective strategy in producing intended effects at a different layer entirely. In other words, these layers have demonstrable cascading effects, and, from a policy perspective, that is something that can be utilized as well.

Scholars have formulated several conceptual schemes for analyzing the Internet—both its policies as well as its operational functionality. One scheme put forth by Milton Mueller is the "three-layer model of assignment." Conceived as a form of technical coordination—specifically in order to hand out unique values and address space and attaching them to users or objects—Mueller divided the Internet into three conceptual layers: (1) a technical layer, where coordination must ensure uniqueness, (2) an economic layer, where decisions are made rationing scarcity, and (3) a policy layer, where decisions are made about rights.[1]

In contrast, Marcus Franda formulated an alternative conceptual scheme. In order to structure his discussion of international regimes and the role they play when it comes to the Internet, Franda conceptualizes such issues in terms of their (1) technical, (2) commercial, and (3) legal dimensions.[2]

Legal scholar Laura DeNardis used the case study of IPv6 to propose four additional areas of Internet governance to supplement the technically focused layers of Milton Mueller: (1) critical Internet resources, (2) intellectual property rights, (3) security, and (4) communication rights.[3]

The seminal Atkins Report written for the NSF in 2003 also used a layer-based model in formulating its now widely used concept of "cyberinfrastructure." Arguing for the development of a new "advanced infrastructure layer," the report cites as its base: (1) integrated electro-optical components of computation, storage, and communication, (2) software programs, services, instruments, data, information, knowledge, and social practices applicable to specific projects, disciplines, and communities of practice, and (3) a cyber-infrastructure layer of enabling hardware, algorithms, software, communications, institutions, and personnel.[4]

While these and numerous other conceptual models have been proposed, the conceptual scheme which will be used here is based on the one originally put forth by Yochai Benkler. In writing to describe the dynamics of media regulation, Benkler conceptualized communications systems into three distinct layers—(1) the physical architecture, (2) the logical infrastructure (or the code), and (3) the content layers.[5] Benkler formulated this three-layer model to explain the dynamics of structural media regulation, arguing that decentralized and democratized information environments are being made possible by emerging network technologies. His layer-based approach was a means of presenting "a new set of regulatory choices" that governments increasingly face in decentralized networked environments.

Benkler's model would later be utilized and developed further by scholars like Lawrence Lessig who took its application beyond the scope of media regulation and toward understanding systems of freedom and control on the Internet, specifically.[6] I argue that Benkler's model is additionally helpful if we extend it even further; toward Internet policymaking. The Internet is, after all, a telecommunications entity, and, as will be demonstrated, the types of policies that apply to regulating its infrastructure, for example, differ fundamentally from those that apply to its content. They are made by different actors and through different processes. No other existing model proves as helpful in explaining the dynamics between competing Internet-related policies.

That does not mean, however, that it is a perfect fit, and upon closer examination it becomes clear that Benkler's three-layer model leaves too much unanswered when transposing it into the Internet policymaking arena for which it was not originally intended. My proposal is, therefore, to build upon Benkler's model in order to make it better suited for the Internet policymaking context which is at the heart of this study. We will accomplish this by redesigning the model to incorporate the following elements: First, the term "architecture" tends to often have a quite different meaning in telecommunications than it does in the context of the Internet and software engineering. Thus, in this study we will substitute the term "infrastructure" in its place, for reasons which will become apparent. Second, this study will demonstrate that Benkler's "code" layer remains too abstract for our purposes here. When identifying the various actors and institutions involved in creating authoritative Internet policies, two fundamentally different types of actors emerge within the code layer, and therefore it is important to draw this distinction in order to formulate a better understanding of governance arrangements. We must emphasize the difference between code, understood as technical protocols, versus code as the software developer's tool for creating applications which the end-user encounters. The result is the emergence of what may ultimately be deemed a fourth layer, separating the single code layer of Benkler into one layer dedicated to technical standards and protocols and another dedicated to software applications. This will highlight not only the substantive differences between institutional actors who either create technical protocols or create private proprietary applications, but also the different *types* of actors involved in decision-making.

Thus, the conceptual model we will employ to understand Internet governance and policymaking will break down the Internet into the following four layers: (1) The Infrastructure, (2) The Technical Protocols, (3) The Software Applications, and (4) The Content.

The purpose of developing this model is to create a lens for policymakers who seek to produce intentional effects, and this is accomplished by breaking down the different political dynamics at each layer so that policymakers' goals can be better aligned with implementation strategies. Toward that end, at each of these four conceptual layers, let us now examine (1) why that layer is important, (2) who governs it, and (3) how policies are being made that affect it.

NOTES

1. Mueller, 17–22.
2. Franda, 13–15.
3. Denardis, 14.
4. Daniel E. Atkins et al., *Revolutionizing Science and Engineering Through Cyberinfrastructure: Report of the National Science Foundation* (January 5, 2003). Retrieved on February 14, 2012 from http://www.nsf.gov/od/oci/reports/atkins.pdf.
5. Yochai Benkler, "From Consumers to Users: Shifting the Deeper Structures of Regulation," *Federal Communications Law Journal* 52 (2000): 561–563.
6. Lawrence Lessig, *The Future of Ideas: The Fate of the Commons in a Connected World* (New York, NY: Vintage Books, 2002).

Chapter 3

Who Governs the Infrastructure?

At its most basic level, the Internet remains a collection of various devices connected to one another. The infrastructure—or the physical wires, cables, and, increasingly, the airwaves—is what actually connects those devices. Without the infrastructure, devices would have no means of communicating with one another, rendering any activity at the other layers meaningless. In sum, the Internet could not exist. As a consequence of this importance, governance over the Internet's infrastructure translates into the ability to make authoritative decisions over Internet access, behavior, and content.

In this chapter, we will examine governance of the infrastructure's two core components: the wired and the wireless. Both have distinct histories which span decades. When it comes to the wired network, it will be illustrated how the Kingsbury Commitment, the "natural monopoly" approach to regulation, and the Telecommunications Act of 1996 all led to the Internet's wired network being governed by an advocacy coalition comprised of a small handful of private telecom firms and cable companies who own and operate the infrastructure, and the national governments around the world that, to varying extents, regulate them. Meanwhile, when it comes to governing the Internet's wireless spectrum, the Communications Act of 1934 and the spectrum-allocation auctions of recent years serve to demonstrate how and why the federal government—primarily the Federal Communications Commission (FCC)—is the central governing authority, along with an epistemic community of engineers that is paramount in guiding its decision-making.

WHY IS THE INFRASTRUCTURE IMPORTANT?

The infrastructure is important because when machines cannot connect to one another, the Internet and all of the communication, information, and transactions it facilitates, cease to exist. Perhaps more than any of the other layers, the infrastructure is also extremely vulnerable. Its physical dimension makes it particularly susceptible to attacks and even outright destruction. It is also characterized as being finite, especially when considering its global scale, and, as a result, several infrastructural "chokepoints" create a scenario where the severing of a few cables may indeed cut off entire continents from the network.

These risks are not merely conjecture, but have, in fact, occurred several times in the Internet's history. For example, in February 2008, five undersea cables that connect Europe to Egypt, and thus the rest of the Middle East all the way to India, were unexpectedly cut almost simultaneously, resulting in Egypt losing 70 percent of its connection to the outside Internet, and nearly 60 percent of India's connectivity was "similarly lost on the westbound route critical to the nation's burgeoning outsourcing industry." Initially, experts "said that ships' anchors, dragged by stormy weather across the sea floor, were the most likely culprit, but Egyptian authorities noted that no ships were in the region."[1]

In a similar case in December 2008, four undersea cables were severed in the same region, and even included some of the very same cables that were involved in the February 2008 incident. Fourteen countries were adversely affected, including The Maldives which were 100 percent down, followed by India, which had an 82 percent disruption. Qatar, Djibouti and the United Arab Emirates were the next most widely affected areas with about 70 percent service interrupted. Disruptions for Saudi Arabia, Egypt and Pakistan range from 51 percent to 55 percent.[2]

These cases simply highlight the fact that cyberspatial activities are still entirely dependent on real-space infrastructure. In each case, individuals' computers still worked, and there were no widespread computer viruses or software problems disrupting the network; it was just that there was no longer a *PHYSICAL* connection between the machines.

The map of the world's Internet undersea cabling infrastructure illustrates that, not only are there a relatively few number of transoceanic cables in total, but certain regions of the world, such as sub-Saharan Africa and New Zealand, are sometimes only connected by a single cable. When the severing of a single cable can cut off large regions of the world from the network entirely, it is evidence of how this major design vulnerability poses a serious cybersecurity threat.

Because the overwhelming majority of Internet users connect through private ISPs, access to the infrastructure is made additionally vulnerable as ISPs have become a "chokepoint" in and of themselves.

This was demonstrated in March 2008, as a business disagreement between two different ISPs escalated to the point where Internet traffic was stopped across parts of the Atlantic. U.S.-based Cogent Communications shut down their links to the Swedish-based ISP Telia in what was described as a contract dispute about the size and locations of the pipes connecting the two ISPs. Many large ISPs—Cogent and Telia, among them—have "peering arrangements" where they agree to interconnect their networks at multiple points and trade roughly equivalent amounts of data traffic. The contract dispute between Cogent and Telia, centered on whether and how to provide fat enough pipes to some of the peering locations, resulted in making it impossible for Swedes and residents of other Nordic and Baltic countries to reach websites hosted on Cogent's network, and vice versa.[3]

The infrastructure is additionally vulnerable, not only through its inter-regional cables or its ISP chokepoints, but also through "Meet-Me rooms." Assuming all of the major undersea and land cables that form the backbone of the Internet's infrastructure are undamaged and operational, there remain, out of necessity, locations where the disparate infrastructural and network elements ultimately connect to each other. These locations, where large networks physically connect to each other, are called "Meet-Me rooms," and they form a chokepoint of their own. If any of the Meet-Me rooms was suddenly damaged or destroyed, again, large regions would be cut off from the rest of the Internet.

For example, the world's most densely populated Meet-Me room is located in the One Wilshire building in downtown Los Angeles. It is a room where over 260 ISPs connect their networks to each other, and, if this facility went down, most of California and parts of the rest of the world would not be able to connect to the Internet. Despite public misperceptions about the Internet being an ultra-high-tech, decentralized system of remote carriers, the Meet-Me room at One Wilshire, characterized as "a phalanx of cabling spill[ing] out of its containers," illustrates both the infrastructure's lack of hardware sophistication as well as its undeniable centralization, particularly at certain physical chokepoints, and in doing so, further highlights its vulnerability.[4]

Ultimately, the infrastructure is vital to the functionality and very existence of the global Internet as the physical connections it facilitates are the prerequisite for any subsequent digital communication to occur.

WHO GOVERNS THE INFRASTRUCTURE?

The physical infrastructure is owned and operated by private commercial firms in the form of telecommunications and cable companies. In some nations, the infrastructure is nationalized.

In terms of governance, the question needs to be framed by determining who has real authority when it comes to making policy decisions over the infrastructure. In other words, despite a confluence of influences, such as the role of the media, the mobilization of grassroots activists, and the lobbying efforts of consumer advocacy groups, the only actors who have decision-making authority over the infrastructure—meaning that they have a proven ability to create policies that constrain or enable behavior with intentional effects—are the owners and operators of that infrastructure. This translates into private telecom firms and cable companies, and the national governments around the world that, to varying extents, regulate them.

The list of telecom firms in the United States which take a leading role in governing the infrastructure primarily include the remaining Baby Bells—AT&T, Verizon, and Qwest—as well as other major backbone players—Level 3 and Sprint Nextel. The entire cable industry put together actually owns very little of the core infrastructure, by comparison. The cable companies' strong influence in policymaking terms is the result of its focus on the notorious "last mile" of the infrastructure.[5]

Increasingly, the wireless spectrum is proving equally vital to the nation's digital infrastructure, and it presents a slightly different governing dynamic because of its transmission of data over the airwaves.

In contrast to the privately owned and operated physical elements of the telecom network, it has long been the policy of the United States that "the public owns the airwaves." The wireless spectrum of frequencies has, therefore, long been regulated by the federal government, in its role as an agent of The People. Because no physical connections need to be constructed in the air, in the same sense as the wires and cables of the physical infrastructure, the ownership of the airwaves has been fundamentally defined in a very different way from that of the physical telecom infrastructure. The wireless spectrum is legally owned by the public, rather than private commercial firms, and the allocation of its frequencies is administered by the government on The People's behalf, creating a situation whereby private companies simply lease access to the spectrum through an application process that occurs on annual basis.

The wireless spectrum in the United States, then, is clearly governed primarily and directly by the federal government. Private commercial firms in the form of broadcasters can be said to govern only in their capacity to make authoritative decisions after they have been granted a license by the federal government, and even at that point, their decisions are subject to further governmental oversight and regulatory policy. This is in stark contrast to the governing dynamic of the physical infrastructure, where the private telecoms and cable companies actually own the infrastructure outright.

For demonstrable evidence that the Internet's infrastructure is governed by national governments and the private telecommunications and cable sectors, one needs only look at the instructive example of cybersecurity policy in the United States.

As concerns over the physical security of important buildings and national landmarks were heightened following the Oklahoma City bombing of the Alfred P. Murrah Federal Building in 1995, President Clinton's Commission on Critical Infrastructure Protection surprised many observers by generating a report titled, "Critical Foundations: Protecting America's Infrastructures,"[6] that "did not focus on the vulnerability of key buildings around the country but instead on the security problems in the new phenomenon of cyberspace."[7]

Cybersecurity policy was, from this relatively early point, viewed in large part as a problem of protecting the physical infrastructure—and the responsibility of creating policies to mitigate that problem was perceived from the outset as falling under the realm of the federal government, working in conjunction with the private telecom sector.

While cybersecurity is a complex case study which will be examined in more detail shortly, for now let us merely assert that cybersecurity policy, while multidimensional, nevertheless contains a strong component highlighting the primacy of infrastructure protection. In other words, as will be demonstrated in the pages to follow, the design of cybersecurity policy itself clearly demonstrates (1) the fundamental importance of the infrastructure, and (2) that the federal government, in conjunction with private telecommunications firms and cable companies, primarily govern that infrastructure.

This is a historically repetitious pattern with established predictability. It was the case through both the Clinton and Bush administrations, and even President Barack Obama, before taking office, proposed that much of his stimulus plan "to rebuild the nation's infrastructure" actually be devoted to, not only roads and bridges, but, additionally, digital infrastructure projects. This is further recognition of the fundamental importance of the Internet infrastructure, and the stimulus' plan to implement such projects primarily through direct government subsidies to private telecom firms and cable companies is evidence that this governing dynamic shows no sign of significantly changing in the immediate future.

HOW ARE POLICIES BEING MADE AT THE INFRASTRUCTURE LAYER?

The pattern of policymaking over the network's physical infrastructure is most clearly an example in support of the advocacy coalitions framework.

In the United States, the history of telecommunications policy forms the basis of the Internet infrastructure's governance. The inventions of the telegraph and telephone in the late nineteenth century meant that a physical network infrastructure must, of necessity, be constructed in order for long-distance communication to occur utilizing these new technologies. Very quickly, the United States government stepped in to ensure that a system of redundancy did not emerge whereby each new entrant into the telecom market would have to build their own network from scratch and continually lay down new wires and cables, identical to those of their business rivals, physically connecting the same locations. This possibility was deemed unnecessarily redundant and grossly inefficient. In response to burgeoning competition, AT&T began acquiring rivals and smaller telephone companies, prompting the federal authorities to initiate antitrust action. In what became known as the "Kingsbury Commitment" in 1913, AT&T agreed to allow competitors to interconnect with its network and pledged that for every new local system acquired, it would sell an equal share of lines to rivals.

The Kingsbury Commitment, then, initiated a policy course on behalf of AT&T and the federal government whereby telephone service would be classified as a "natural monopoly." This argument presumed that redundant telephone infrastructure was economically inefficient, and that monopoly power could simply be mitigated through rate regulation.[8] Support was ensured for this policy because AT&T would enjoy government protection from competition (limiting access to the market), while public officials could guarantee to their constituents "One Policy, One System, Universal Service."[9] Regulatory authority was formalized in the Mann-Elkins Act of 1910 as vested in the Interstate Commerce Commission (ICC), and later in the FCC in 1934, justified on the basis of best serving the "public interest, convenience, or necessity."[10]

This natural monopoly argument, coupled with a policy context whereby only one network infrastructure would be built and operated by the private sector, but regulated by the federal government, would set a course of path dependency that would endure for most of the next century.

In the 1970s, the U.S. Justice Department filed an antitrust lawsuit against AT&T based on complaints of anticompetitive practices by MCI and other long-distance service providers, prompting a landmark settlement which forced AT&T to restrict its services to the long-distance telecom market, while its local services were divested into seven regional operating companies, which after a series of mergers and acquisitions later became four—SBC, Verizon, Bellsouth, and Qwest. Thus began in 1982 a policy environment where local and long-distance services would constitute a "regulatory disconnect," where the federal government imposed an artificial legal distinction between local and long-distance services, despite technological advances that would render any such distinction obsolete.[11] Once again, during this era of the "Baby Bells," a sudden and transformative shift in the telecom policy environment was followed by years of relative stability and entrenched interests.

Then, in the early and mid-1990s, with the advent of the World Wide Web and other digital network technologies, regulatory challenges created by the local versus long-distance distinction began to cross industry lines. The new telecom environment became one of "convergence," meaning that digital technologies could allow operators to offer all types of telecommunication services—local, long distance, wireless telephony, cable broadcast, and Internet access—regardless of which type of network infrastructure they used. Consequently, new regulatory problems emerged such as whether the Internet should be classified as long distance, whether cellular service could be classified as local, or whether the Baby Bells could provide "information services" to clients with interregional offices.

In response to these external technological forces that were fundamentally altering the regulatory landscape, Congress enacted the Telecommunications Act of 1996, which sought to resolve these disputes by putting an end to the monopoly franchise system governing local calling,[12] thus beginning the contemporary policy environment of "deregulation."

The Telecommunications Act of 1996 was nothing less than a dramatic reformulation of the entire nation's communications policy, which had been essentially unchanged since 1934.

What forces led to the complete reversal of nearly 70 years of telecommunications policy? It is crucial to note at this point that, at a fundamental level, the telecommunications sector is characteristically different than some other policy arenas because of its heavy dependency on technical and scientific expertise. Because of this, the political process is only one element in the formulation of policy, and technical decisions often have the effect of law. The integral reliance of technical expertise in telecommunications has historically given the epistemic community of engineers a disproportionally large role in formulating telecom policy.[13]

In political terms, however, it was the confluence of three events which led to the Telecommunications Act. First, the Clinton administration, and particularly Vice President Al Gore, came to office determined to "wire the nation's classrooms," and to do so through market competition rather than public spending.[14] Several Democratic legislative efforts that strongly reflected the Administration's position failed in 1993, such as H.R. 3636 sponsored by Ed Markey (D-MA) and Jack Fields (R-TX), finding opposition to principles of guaranteed universal service, mandatory price breaks to specialized entities ranging from schools to rural hospitals, and stringent limitations on the Bells' entry into long-distance markets.

Second, the political environment was then dramatically reshaped in 1994 with the Republican ascendancy in Congress. While raising opposition to numerous Clinton administration proposals, their determination to limit the size and scope of government manifested itself into a push for deregulation aimed at the FCC. Deregulation of the telecom sector, it was argued, called for the elimination of regulatory distinctions between local, long-distance, cable, wireless, and data communication. This would allow the concentration of broadcasting companies in a single media market, as well as permit telecom providers to begin offering services into alternative markets (such as local phone companies being able to offer long-distance services) with minimal government intervention.[15]

Third, the World Wide Web and the first browsers were simultaneously being introduced and reaching a critical mass of users within the United States. Other new digital network technologies were discovered that changed the context of the regulatory debate. For example, the engineering community had discovered that by sending packets of digital information, coded in bits in the frequency of the radio waves transmitted over the air, they could communicate with much more information than in analog broadcast. The big point of the crucial new discovery was that by digitizing and compressing the signals, over a single channel's worth of spectrum the broadcasters could deliver not just one but a total of six simultaneous programs in standard definition. This meant that cable television companies could now enter the markets for local and long-distance telephony as well, using a separate network infrastructure, and introducing them as viable competitors in the marketplace. Consequently, federal rate regulation could be rescinded, and competition from these digital broadcasters could keep process costs down.[16]

The result was a unique policy window created by the confluence of these three events—the Clinton administration's desire to wire the nation's classrooms, the Republicans' push for deregulation, and the rise of network technologies that fundamentally altered the technical possibilities in the telecom environment. New solutions and possibilities presented themselves in these digital technologies, and there was sufficient political will on both sides of the aisle to push for change in a similar direction.

In order to additionally understand the process of how the Telecommunications Act came into being, it is also useful to examine the advocacy coalitions participating in the debate. For several years prior to enactment, the telecom debate centered primarily on the Bell companies who after nearly a century of regulated monopoly now wanted the freedom to offer a broad range of services from telecom equipment to local and long-distance services to cable television programming to data delivery geared toward the Internet.

These Bell companies were opposed by long-distance firms such as AT&T and WorldCom who shared an interest in being able to rent phone capacity at wholesale prices, or in other words, to gain access to the Bell's local networks. Meanwhile, cable operators wanted an end to price regulation, promising in return to become the "second wire into the home" that would abolish cross-ownership limitations for telephony and broadcast. Also, broadcasters wanted to abolish concentration of ownership and cross-ownership restrictions, for both television and radio, and argued for free and exclusive access to the adjoining spectrum for digital uses, especially since the auction model had become prominent in spectrum management.

Additionally, computing interests joined the anti-Bell forces in wanting to protect from Bell domination new Internet businesses that use the communications networks, and opposed classifying Internet services as within the realm of telecommunications. Public interest groups and nonprofit organizations also emerged opposing the concentration of ownership (of both infrastructure and content), the relaxation of public trusteeship regulation such as licensing oversight, and the removal of price regulation. It is crucial to note, however, that while these groups opposed the Bell companies on the various fronts mentioned, there was a general consensus by all parties that the system should be geared toward a more deregulated, liberalized environment. They only differed on the extent to which deregulation should reach.[17]

All of the aforementioned political and business interests, the Bells included, collectively created a decidedly pro-liberalization, pro-competition environment within the telecom sector. It was this alignment of influential actors and groups, both within and outside of the political system, which ultimately led to policy change within this particular policy subsystem.

The passage of the Telecommunications Act, then, is most clearly an example in support of the advocacy coalitions framework, as prescribed by Sabatier and Jenkins-Smith,[18] insofar as (1) technical expertise clearly plays an important role concerning the magnitude and facets of both the causes and probable impacts of various solutions in the sector, (2) a telecom subsystem clearly exists with public and private actors actively concerned with the regulatory issue over an extended period of time, and (3) that the Telecommunications Act incorporates implicit theories on how best to achieve its objectives of wiring the nation's classrooms and eliminating artificial legal distinctions between telecom domains by conceptualizing the issue in much the same way as belief systems, implementing a theoretical policy shift involving value priorities and conceptions of various policy instruments, through the Act's ultimate emphasis on deregulation and market-based solutions to telecommunications.

The fact that U.S. telecommunications policy went virtually unchanged for 70 years before its dramatic reformulation in 1996 might seemingly support the framework of Punctuated Equilibrium, which argues that the political system displays institutionally enforced stability that is punctuated by brief periods of volatile change.[19] However, Baumgartner and Jones fail to account for a unique occurrence in the process behind the Telecommunications Act—that it was actually the beneficiaries of the path dependency who advocated for change. Rather than witnessing a sudden mobilization of bias[20] where the scope of the debate is enlarged to include various new actors in the policy process, the telecommunications arena remained primarily influenced by the same large corporate firms which had been the beneficiaries of the previous natural monopoly and subsequent highly regulated telecom systems for decades. Instead, the dramatic policy change was primarily the result of innovative technologies and the new possibilities they offered to entrenched firms in the subsystem, leading to those same firms, acting in their rational self-interest, to demand the elimination of protective regulations because technology offered more promising benefits.

Multiple Streams theory might also seemingly be used to explain the passage of the Telecommunications Act, in that it was the convergence of the policy and political streams—or, in other words, the existence of reasonable proposals and alternatives coupled with strong political will on behalf of top influential actors—which made policy change possible.[21] However, Kingdon fails to account for the fact that, in formulating this telecom policy, there was a complete absence of any viable problem stream. No pressing problem found on the national agenda was demanding attention in telecommunications, and surely there was no public outcry with the previously existing system. Rather, once again, it was technology which provided, not a problem to be dealt with, but new possibilities which policy entrepreneurs sought to deliver to various advocacy Groups.

In contrast to the case of the network's physical infrastructure, the pattern of policymaking over the wireless spectrum suggests a slightly different governing dynamic, where an epistemic communities approach, specifically regarding engineers, is paramount.

In concrete policymaking terms, decisions over how the government allocates wireless frequencies are illustrative of how policies get made at this area of the Infrastructure layer. Ever since it was determined that the public owns the airwaves, efforts have been made to determine how best to allocate frequencies—to whom, and based on what criteria.

The Radio Act of 1927 was passed to provide temporary regulation to correct an immediate problem of broadcasters failing to respect the frequencies used by their rivals. It established the five member Federal Radio Commission (FRC) and granted it broad powers to bring order to the airwaves by reducing the total number of broadcasters so that the remaining stations would be able to broadcast more effectively—solving the technical dilemma of multiple broadcasters sharing the same frequency. The Radio Act did not provide specific guidelines for the FRC to use in selecting broadcasters for the limited number of frequencies, but rather called for the allocation of licenses on the basis of who best served "*the public interest, convenience, or necessity*," a phrase of great importance adopted from public utilities law.[22]

It was at this stage where the epistemic community of scientific engineers would play the central role in creating the future structure of the U.S. broadcasting system. The FRC immediately convened hearings in March and April 1927 to listen to suggestions from broadcasters on how the FRC could best regulate the medium. The agenda for the hearings was structured around engineering concerns and the sessions were dominated by the testimony of corporate-affiliated radio engineers.[23] Given the emphasis on engineering and technical criteria for reallocating the broadcast spectrum, the hearings were largely devoid of controversy, in spite of the press and members of Congress being invited to attend.[24]

The FRC announced its reallocation plan in August 1928. In what is known as General Order 40, the FRC acknowledged that Congress had given it no indication as to how best to determine the meaning of what best served "the public interest, convenience, or necessity," and so the statement asserted that the FRC would interpret the phrase as meaning that the FRC should strive "to bring about the best possible broadcasting reception conditions throughout the United States."[25] Consequently, the FRC would rely on engineers to provide technical expertise in bringing about optimal reception conditions.

Ultimately, the FRC's actions were defended by Congressional committee members such as Orestes H. Caldwell as being made in the best interests of the listeners. Caldwell, a trained electrical engineer, was outspoken in his view that it would undoubtedly best serve the public interest to "extend the number of radio listeners until we put a set in every home."[26]

Engineers acted as vital agents in the policy process outlined above, insofar as they were the primary consultants and advisors in establishing the technical criteria by which the allocation of broadcasting frequencies would depend. They were an outsider, nongovernmental group who greatly influenced the Executive branch of the federal government.

The central idea that acted as the episteme in this case would be the engineers' continued assertion that the public interest would be best served by bringing about the best possible broadcasting reception conditions throughout the United States, and this would be most effectively accomplished by assigning frequency rights to those broadcasters with the capability to provide the best technical equipment, and with high levels of capitalization to maintain and upgrade that equipment.

Epistemic communities tend to have a shared set of symbols, references, and mutual expectations, and in the scientific community, these tend to be any policies, events, or actions that help to further the advancement of science. Ruggie's notion of implementing standards of "normal" behavior[27] is evident in the engineers' unchallenged analysis of using technical standards to determine which broadcasting stations would be deemed worthy of frequency rights.

After the passage of the Radio Act in 1927, there was, as Secretary Herbert Hoover remarked, a pressing need to "clear up the chaos of interference and howls in radio reception."[28] Such a problem was technical by nature, as it was caused by a limited number of frequencies on the radio wave spectrum comparable to the number of prospective broadcasters. The FRC hearings that occurred in March and April 1927 which would form the policy basis for General Order 40 were themselves structured around engineering concerns and the sessions dominated by the testimony of radio engineers. This is a clear case of policymakers deferring to the expertise of nongovernmental actors, whose knowledge of the technical subject they did not deem themselves as qualified to challenge.

The FRC allocating committee led by commissioners Caldwell and Sam Pickard was explicitly mandated by Congress to "consult with experts," and would later regard reallocation as "strictly an engineering problem." Additionally, "clear channels" were developed solely based on the recommendation of the engineering group advising the FRC, as the epistemic community sought its goal of better telecommunications infrastructure on a national level. The influence that engineers held over the entire policy process was tremendous, as policymakers, recognizing a highly technical environment where they had little understanding of how things worked, to a large extent took it for granted the need to defer to the expertise of the epistemic community.

As a result, the eventual broadcasting policy—the seminal Communications Act of 1934—was a policy outcome clearly consistent with those ideas espoused by the epistemic community. Even beginning with General Order 40, the idea that the broadcast spectrum should be distributed according to applicant stations' relative ability to provide high-quality reception conditions on a national level became from the early stages of the policy process the guiding doctrine of how to define "the public interest, convenience, or necessity." The Communications Act of 1934 codified this still guiding principle of U.S. broadcasting policy and also laid the foundation for the process by which stations compete for frequencies based on their technical capacity, not necessarily their broadcasted content, and the standard resting on the epistemic community's underlying belief in the best and most modern telecommunications equipment which forms the entire system's infrastructure. The most direct result evident from this episteme in policy is the decades-long entrenchment of the commercial, national networks dominating the medium with effectively unchallenged status.

The same pattern of policymaking exists today as it applies to the governance of the wireless spectrum for Internet usage. The basic principle of the public owning the airwaves remains in place, and the federal government, through the FCC, continues to allocate broadcast frequencies on the basis of recommendations made by the epistemic community of engineers.

For example, in March 2008, the FCC held an auction for the rights to the much-coveted 700 MHz band of the wireless spectrum. The auction for these licenses raised over $19 billion, and the major participants included the likes of traditional telecom firms like AT&T and Verizon, as well as Internet service companies like Google. During the process of this auction, and at the urging of the engineering community, the FCC forced the major telecom firms to open their wireless networks to a broader array of telephone equipment and Internet applications.[29]

Aside from the issue of allocating licenses on the 700 MHz band, the FCC has also taken on the role of deciding how to govern the so-called "white spaces"—unused portions of the wireless spectrum that exist between UHF channels. Despite the protests of traditional telecom firms like AT&T, Verizon, and Comcast, on November 4, 2008, the FCC voted 5–0 to approve the unlicensed use of white space.[30] This decision came about as the result of a strong movement among the engineering community calling for free, unregulated spectrum that could be used, like Wi-Fi Internet access, to create new technologies and new markets.[31]

This demonstrates, yet again, how the federal government, and particularly the FCC, play the primary role in governing the wireless spectrum, while the epistemic community of engineers, due to their technical expertise, maintain a strong authoritative role in guiding the government's policymaking process.

In summary, as the Internet and digital telecommunications continue to become more integrated into the nation's economy and culture, the physical infrastructure that makes it possible only gains in relevance, and its problems of vulnerability only become more consequential. The Internet's infrastructure can be conceptually divided up into two components: (1) its physical infrastructure, which is governed by the private telecom firms and cable companies who own it, and the national governments that regulate them, and where an advocacy coalitions framework best describes its pattern of policymaking; and (2) its wireless spectrum, which is governed by the federal government through the FCC, and where an epistemic communities approach best characterizes how policy decisions are made governing the allocation of frequencies.

NOTES

1. John Borland, "Analyzing the Internet Collapse," *MIT Technology Review* February 5, 2008. Retrieved on January 21, 2009 from http://www.technologyreview.com/Infotech/20152/?a=f.

2. Kim Zetter, "Undersea Cables Cut; 14 Countries Lose Web—Updated," *Wired* December 19, 2008. Retrieved on January 21, 2009 from http://blog.wired.com/27bstroke6/2008/12/mediterranean-c.html.

3. Ryan Singel, "ISP Quarrel Partitions Internet," *Wired* March 18, 2008. Retrieved on January 21, 2009 from http://blog.wired.com/27bstroke6/2008/03/isp-quarrel-par.html.

4. Dave Bullock, "A Lesson in Internet Anatomy: The World's Densest Meet-Me Room," *Wired* March 3, 2008. Retrieved on January 21, 2009 from http://www.wired.com/techbiz/it/multimedia/2008/03/gallery_one_wilshire.

5. "Who Owns the Internet? We Have a Map That Shows You," *CIO*. Retrieved on February 10, 2009 from http://advice.cio.com/node/209.

6. *Critical Foundations: Protecting America's Infrastructures* October 1997. Retrieved on November 11, 2008 from http://chnm.gmu.edu/cipdigitalarchive/files/5_CriticalFoundationsPCCIP.pdf.

7. Richard A. Clarke, *Your Government Failed You: Breaking the Cycle of National Security Disasters* (New York, NY: Ecco, 2008), 289.

8. Diane Katz and Theodore Bolema, *Crossed Lines: Regulatory Missteps in Telecommunications Policy* (Mackinac Center Report, 2003).

9. Theodore Newton Vail (President), *AT&T Corp* December 19, 1913.

10. Robert W. McChesney, *Telecommunications, Mass Media, and Democracy: The Battle for the Control of U.S. Broadcasting, 1928–1935* (New York, NY: Oxford University Press, 1993).

11. Katz and Bolema.

12. Katz and Bolema.

13. McChesney.

14. Reed E. Hundt, *You Say You Want A Revolution: A Story of Information Age Politics* (New Haven, CT: Yale University Press, 2000).

15. Patricia Aufderheide, *Communications Policy and the Public Interest: The Telecommunications Act of 1996* (New York, NY: Guilford Press, 1999).

16. Aufderheide.

17. Aufderheide.

18. Paul A. Sabatier and Hank C. Jenkins-Smith, "The Advocacy Coalition Framework: An Assessment," *Theories of the Policy Process*, ed. Paul A. Sabatier (Boulder, CO: Westview Press, 1999).

19. James L. True, Bryan D. Jones and Frank R. Baumgartner, "Punctuated Equilibrium Theory: Explaining Stability and Change in American Policymaking," *Theories of the Policy Process*, ed. Paul A. Sabatier (Boulder, CO: Westview Press, 1999).

20. E. E. Schattschneider, *The Semi-Sovereign People: A Realist's View of Democracy in America* (New York, NY: Holt, Reinhart, and Winston, 1960).

21. John W. Kingdon, *Agendas, Alternatives, and Public Policies*, 2nd ed. (New York, NY: Addison-Wesley Educational Publishers Inc., (1995).

22. McChesney.

23. McChesney.

24. *Memorandum: Allocation Broadcasting Channels to Zones and States*, FCC General Correspondence, March 30, 1928.

25. "FRC Interpretation of Public Interest," *Documents of American Broadcasting*, ed. Frank J. Kahn (Englewood Cliffs, NJ: Prentice-Hall, 1984).

26. McChesney.

27. McChesney.

28. *Statement by Secretary Herbert Hoover Regarding the Radio Situation*, HH-C Radio Correspondence, February 24, 1927.

29. Stephen Labaton, "Wireless Spectrum Auction Raises $19 Billion," *New York Times* March 19, 2008. Retrieved on February 27, 2009 from http://www.nytimes.com/2008/03/19/technology/19fcc.html?_r=1&oref=slogin.

30. "FCC Adopts Rules for Unlicensed Use of Television White Spaces," *Federal Communications Commission News* November 4, 2008. Retrieved on February 27, 2009 from http://hraunfoss.fcc.gov/edocs_public/attachmatch/DOC-286566A1.doc.

31. Priya Ganapati, "FCC White Spaces Decision Kicks Off the Next Wireless Revolution," *Wired* November 5, 2008. Retrieved on February 27, 2009 from http://blog.wired.com/gadgets/2008/11/fccs-decision-t.html.

Chapter 4

Who Governs the Technical Protocols?

The reason why the Internet has often proved challenging to govern, regulate, or control is that the network itself is "dumb."[1] The technical protocols that allow digital communication to occur over it were intentionally designed to ignore the content of transmissions and instead focus only on the engineering goal of efficient routing. To be sure, this was the result of an explicit policy decision, and it is why content, to such a large degree, still proliferates so freely in cyberspace. Thus, the point needs to be emphasized: The extent to which Internet content is regulable or not regulable is largely determined by decisions which are built into the code of the network itself.

In this chapter, we will examine how technical decisions over protocols have inherently political consequences. A study of the Internet's TCP/IP suite as well as other prominent protocols that ensure the very functionality of the Web itself will demonstrate that decision-making authority is held by a small handful of international engineering consortium groups—primarily, the ISOC, its IETF, and the W3C—and we will then analyze the constitutional makeup of these organizations. By considering cases such as the Internet-OSI Standards War and, more recently, the implementation of IPv6, we will argue that the decision-making process within these institutions is best characterized by the "rough consensus principle," which is fundamentally open and transparent. Finally, we assert that the decisions over which technical protocols to adopt, and how they are to be designed, are, in themselves, an important form of policy that constrain and enable behavior on the Internet.

WHY ARE THE PROTOCOLS IMPORTANT?

The technical standards and protocols underlying the Internet are the languages by which digital communication occurs between different devices. While an infrastructure is first required to physically connect separate devices, once the connection is in place, the machines require a common language in order to communicate with one another.

The basic TCP/IP suite of protocols forms the basis of the modern Internet. TCP/IP allows for the sending and receiving of packets of data. Two pieces of information are required for an exchange of data to be successful—the IP address of the machine from where the data is being sent, and the IP address of the machine which is its destination.

The protocols, however, do not reveal much information about the nature of the data being sent—and this is a crucial point. Because IP addresses are virtual, TCP/IP does not actually reveal who is sending the data, nor where they are sending it from in a geographical sense. The system has been designed not to care about what type of data it is, nor its purpose, but rather simply to include the minimum information necessary to successfully exchange the data from one machine to another. From the perspective of the network, any additional information is considered "an unnecessary surplus."[2]

Thus, TCP/IP was designed according to the principle of minimalism, with the goal being optimal network efficiency. The protocols are the result of a decision-making process which was characteristically focused on engineering concerns.

However, in being designed as such, the protocols inevitably led to a wide assortment of political, social, and economic consequences. As described by legal scholar Laura DeNardis, "protocols, while often established primarily by private actors, are intertwined with socioeconomic and political order."[3] While the adoption of TCP/IP, and its noninclusion of any information other than what was minimally necessary for successful transmission, may have led to efficient engineering, it also simultaneously was a decision to disable other potential forms of control. For example, because TCP/IP neglects to collect any information on the sender other than the sending machine's IP address, this makes it exceedingly more difficult for governments to identify the human sources of criminal behavior online. Various other forms of control or regulatory oversight are similarly rendered difficult, if not impossible, based on the relative anonymity that TCP/IP fosters.

This is why we argue that technical decisions have political consequences. The open design of TCP/IP, HTTP, CGI, and other standards and protocols as well, directly affect governments' capabilities when it comes to the regulation of speech, the protection of privacy, the distribution of copyrighted material, the identification and prosecution of criminals, and much more. The relative openness and anonymity that cyberspace provides is built into the code of the Internet itself, through the design of its protocols. If protocols were designed and adopted to, conversely, limit that openness, which is certainly a technical possibility, it would have dramatic consequences on all of the political issues just mentioned. Ultimately, since the design of these protocols has such significant effects, it is crucial to remember that they have been designed and adopted by actual people, making intentional decisions to do so, and thus determining who these people are, and what guides their decision-making is of paramount importance.

There are many other technical standards and protocols that enable the Internet and the World Wide Web to function aside from TCP/IP. Most prominent among these include HTML—which is the elemental markup language by which all web pages are displayed through a browser. Additionally, XML, CSS, SMIL, XSLT, CGI, DOM, SOAP, and hundreds of others all play a role in shaping the Internet environment in technical terms. Again, as with TCP/IP, decisions over which protocols to adopt, and their final form, directly determine the content of what is available, as well as the behavior of people, online. These standards and protocols directly enable certain forms of action, while making others technically impossible.

The protocols are important, then, because they are, indeed, a binding form of policy. Technical decisions over which protocols to adopt set the rules for behavior, and thus lead to inherently political consequences. The protocols are what shape the network to be either more regulable, or less regulable, and their ultimate effect is that they determine both what types of behaviors are possible online, as well as what types of policies seeking to influence those behaviors will be meaningful and, or, effective.

WHO GOVERNS THE PROTOCOLS?

The major Internet technical standards and protocols are governed directly by several international engineering consortium groups. Primarily, these are the ISOC, its IETF, and the W3C.

The ISOC is the organizational home of the Internet Architecture Board (IAB) and the IETF, which handles much of the open development of the Internet's behind-the-scenes architectural issues. It is an international nonprofit organization founded in 1992 to provide leadership in Internet-related standards, education, and policy. It also has more than 90 organizational members and more than 26,000 individual members in over 80 chapters around the world, with its main offices in Washington D.C. and Geneva, Switzerland. The membership is comprised of commercial companies, government agencies, and foundations that have historically been at the forefront of developing the Internet and its technologies, as well as new innovative and entrepreneurial organizations contributing to maintain that dynamic.[4]

ISOC's members are distributed throughout the world. Fully, 27 percent come from North America, another 27 percent from Asia, 23 percent from Europe, 12 percent from Africa, 6 percent from Latin America and the Caribbean, and 5 percent from Oceania.[5]

Furthermore, the distribution of ISOC members can be classified by industry type. Currently, there are 6 members that are "educational institutions," 28 "product providers," 17 "NICs, Registrars, and IP Registries," 2 "financial institutions," 18 "network access providers," 16 "organizations for research, professions, industries, and standards," 5 "government agencies," and 5 "uncategorized."[6]

As with any formal institution, the organizational structure significantly affects processes and outcomes. ISOC has 6 membership levels; each of which has its own financial requirements to join, and grants corresponding powers to its members. There are currently 6 Platinum members (the highest ranking level), 6 Gold members, 6 Silver members 39 Executive members, 23 Professional members, and 12 Small Business members. The amount

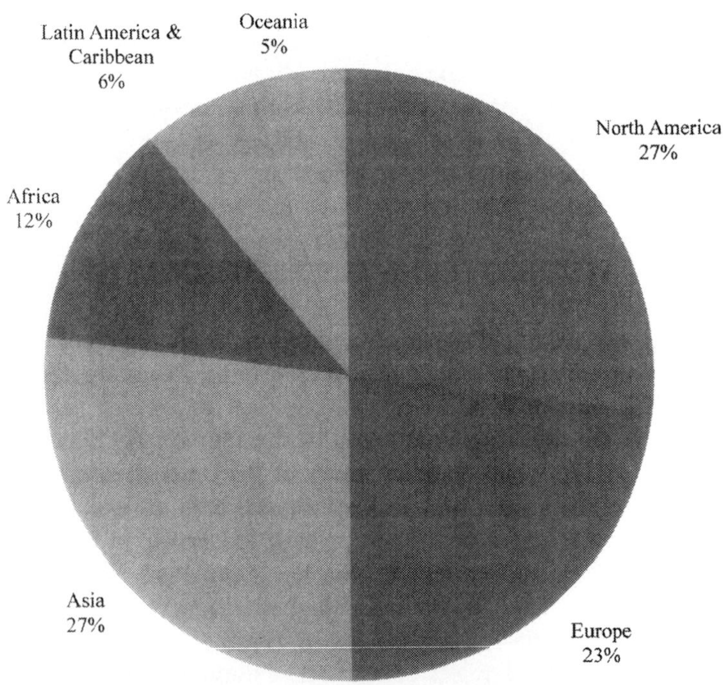

Figure 4.1 ISOC Membership by Geographic Region

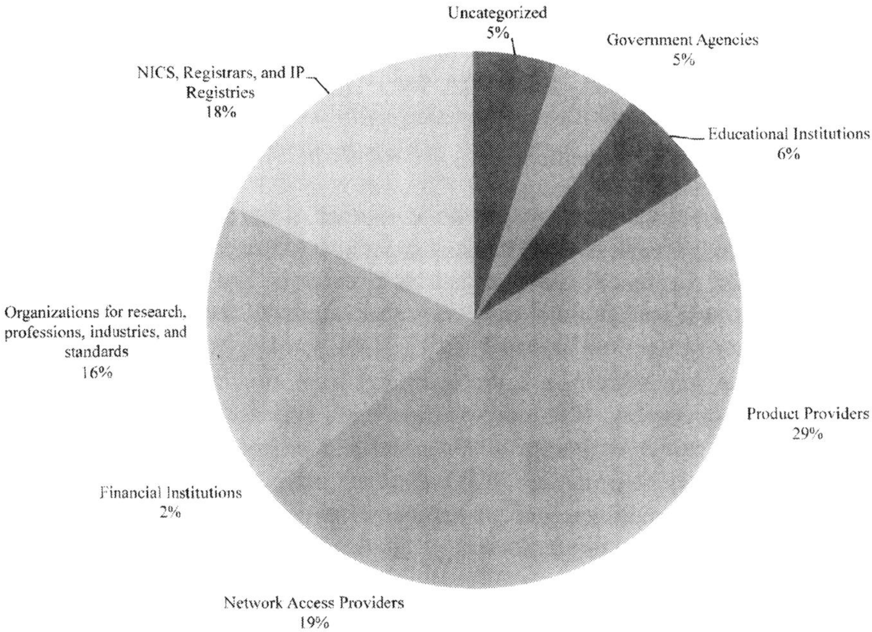

Figure 4.2 ISOC Membership by Organizational Type

required for an organization to join ISOC ranges from $100,000 annually for Platinum members to $1,250 annually for nonprofit Small Business members. In exchange, the powers granted to members according to level involve having greater influence over designating funds for specific Internet activities and projects, as well as rights to use the ISOC logo and gain more prominent acknowledgments in ISOC web pages and publications.[7]

The IETF is responsible for overseeing how TCP/IP protocols evolve. It is an open international community of network designers, operators, vendors, and researchers. As such, it has no formal membership or membership requirements, relying instead completely on volunteers and "open to any interested individual,"[8] though their work is usually funded by their employers or sponsors; for example, the current chairperson, Russ Housley, is funded by VeriSign and the U.S. government's NSA.[9] Its first meeting took place in January 1986 and consisted of 21 government-funded researchers. As of 2008, meetings now attain attendance of approximately 1100 people per meeting.[10] However, in the early 1990s, the IETF changed its institutional form from an activity of the U.S. government to an independent, international activity associated with the ISOC. Because the IETF has no formal members, funding, as well as the legal framework for the activities of the IETF, also comes from ISOC.

For clarification, while the IETF predates the establishment of ISOC, and indeed ISOC grew out of the IETF in order to support those functions that require a corporate form rather than the IETF's simple ad hoc approach, ISOC nevertheless is now the parent corporation of the IETF. For example, all IETF "Request for Comments" (RFC) documents are copyrighted by ISOC.

The W3C develops standards for the evolution of the most popular part of the Internet, the World Wide Web. It is an industry consortium run by MIT's Laboratory for Computer Science and its director is Tim Berners-Lee—the inventor of the World Wide Web.[11] Its list of standards for which it can take credit include most prominently HTML, XML, and CSS, but include over 100 others as well which form the technical basis for the Web that people interact with every day. It is a consortium with over 400 member organizations, many of which maintain full-time staffs of technical experts dedicated to the standardization process.[12] W3C members include businesses, nonprofit organizations, universities, and governmental entities, however, there is no provision for individual membership.[13] It has headquarters in the United States, France, and Japan, as well as regional offices in Australia, the Netherlands, Luxemburg, Belgium, China, Finland, Germany, Austria, Greece, Hungary, India, Ireland, Israel, Italy, Japan, South Korea, Morocco, South Africa, Spain, Sweden, and the United Kingdom. The W3C operations are financially supported by a combination of member dues, research grants, and other sources of public and private funding.[14]

Membership in the W3C is contingent on fees paid. In their stated attempt to "promote a diverse membership that represents the interests of organizations around the world," W3C fees vary for each member and are determined by a "Membership Fee Calculator" based on an algorithm that takes into account the applicant organization's annual revenues, type, and location of headquarters. Thus, for example, a small company in India would pay $953 annually, while a nonprofit in the United States would pay $6,350, and a large company in France would pay 65,000 euros.[15]

Certainly, other standards-setting organizations play significant roles in governing the Protocol layer as well. For instance, the Institute for Electrical and Electronics Engineers (IEEE) is responsible for the 802.11 standard, commonly known as Wi-Fi, which carries out wireless local area network (WLAN) computer communications in the 2.4, 3.6 and 5 GHz frequency bands, and has rapidly become adopted by the mainstream public.

When it comes to who has decision-making authority over technical protocols—and, thus, make decisions that will affect what content is ultimately available as well as how people may behaviorally interact with it—these three organizations—ISOC, the IETF, and the W3C—play the most direct and prominent role in governing the development and implementation of policies.

HOW ARE POLICIES MADE AT THE PROTOCOLS LAYER?

Policies regarding the development and implementation of technical standards and protocols are made by the aforementioned organizations primarily through a process known as the "rough consensus principle."

Rough consensus has been the guiding governance principle throughout the Internet's developmental history. It has been the de facto method of generating norms and usable standards in an environment where there has been a historical lack of central regulatory agencies. Rough consensus overcomes initial problems of legitimacy and fragmented levels of adoption and acceptance through a process of continuous testing and refinement of proposals that are ultimately measured by responsive models of compliance. This framework has even been applied in academic circles to such fields as transnational lawmaking, where the public and private activities of transnational actors often takes place in the absence of traditional enforcement mechanisms.[16]

The origin of the phrase stems from the now-infamous quote by one of the Internet's pioneers, David Clark, at an IETF meeting in 1992: "We reject: kings, presidents, and voting. We believe in: rough consensus and running code."[17]

The IETF describes its process of achieving rough consensus as the following:

> Working groups make decisions through a "rough consensus" process. IETF consensus does not require that all participants agree although this is, of course, preferred. In general, the dominant view of the working group shall prevail. (However, it must be noted that "dominance" is not to be determined on the basis of volume or persistence, but rather a more general sense of agreement). Consensus can be determined by a show of hands, humming, or any other means on which the WG agrees (by rough consensus, of course). Note that 51% of the working group does not qualify as "rough consensus" and 99% is better than rough. It is up to the Chair to determine if rough consensus has been reached.[18]

To clarify, because the IETF and similar groups never had members, only volunteer participants, having a formal voting structure was viewed as problematic. The rough consensus principle includes the ideas that newcomers are encouraged to contribute their expertise, and working-group leaders ought to approve proposals that enjoy broad support within the group. An acceptable level of agreement can usually be placed at around 80–90 percent—"a level high enough to demonstrate strong support, but flexible enough to work in the absence of unanimity." In short, rough consensus is "an informal process in which a proposal must answer to criticisms, but need not be held up if supported by a vast majority of the group."[19]

"Rough Consensus and Running Code" is a phrase that, as Andrew Russell has argued, captures both the technical and the political values of Internet engineers—during its early history leading up to, and including, the present day.[20] It has been additionally described as, "a manifesto that will define our generation."[21]

The reason for such hyperbole is the consideration of alternative methods that may have come to define the Internet standardization process if the rough consensus principle had not come about. In other words, it is crucial to note that rough consensus was not inevitable, nor was it deterministic. During the 1970s, it seemed quite likely that formal standards-setting institutions, such as the International Telecommunication Union (ITU) and the International Organization for Standardization (ISO), would ultimately create the Internet's standards just as they had for those of telecommunications; that is, in a top-down manner, and stemming from a centralized authority.

In fact, a rival process for setting standardization policies had been introduced in 1977. The long-established ISO attempted to define its own vision of a network architecture known as Open Systems Interconnection (OSI). Despite the OSI seven-layer model being endorsed by national governments and computer science departments around the world, it failed to adjust to the rapidly changing technological environment in network computing. Rival systems like TCP/IP, which were being developed by volunteer researchers in the newly formed groups such as the IETF and IAB, proved more quick to adapt. Rough consensus became the guiding principle for these groups out of practical necessity; the groups were volunteer-based and did not have formal members, they were open to anyone willing to participate, and meeting attendance was not required for a collection of individuals loosely scattered across the globe. Thus, the malleable nature of the participants in the decision-making process brought about a system of policymaking whereby rough consensus became the only effective means of collaboration.[22]

As a result, the informal process of achieving rough consensus, used by the IETF in developing TCP/IP, stood in stark contrast to the bureaucratic-political approach that characterized OSI. This is why the rough consensus principle aptly describes not only the technical process of standardization, but also, significantly, the institutional system that governed it.

This so-called Internet-OSI Standards War was framed in terms of everything from a "constitutional crisis" to a "religious war." William Drake summarized the issue as follows:

> The debate is not merely about the efficacy of two sets of standards, but it is rather between two competing visions of how international standardization processes and network development should be organized and controlled.[23]

Eventually, TCP/IP pushed the OSI model into the background. The IETF's protocols were practical and, because they were the result of intense implementation discussion and testing, they actually worked, whereas ISO committees produced what came to be viewed as theoretical models that were difficult to alter or fully implement.[24] One expression at the time encapsulated the situation, "OSI is a beautiful dream, and TCP/IP is living it."[25] Rough consensus had proven to be a more effective method for creating standardization policies than had a traditional top-down approach.

Thus, rough consensus became the mechanism by which policies are made regarding Internet standards and protocols, and it has maintained that status through the present day in all of the aforementioned dominant governing institutions—ISOC, the IETF, and the W3C.

Achieving rough consensus in decision-making is often a heavily politicized process as well, as demonstrated with the case of the IPv6 protocol. Recognizing that the finite supply of Internet addresses would eventually reach its limit, the IETF designed a new system—IPv6—to replace the existing system—IPv4—expanding the pool of potential IP addresses from 4.3 billion to 340 undecillion. Here, too, the process that ultimately led to the adoption of IPv6 was ripe with conflict and competing interests. As Laura DeNardis described, it "involved complex technical choices, controversial decisions, competition among information technology companies, resistance from large American companies . . . and an institutional choice between a protocol developed within the prevailing Internet governance institutions and one promoted by a more international institution."[26] Over a decade since its design, IPv6 deployment remains largely unadopted.

How does the rough consensus principle drive the decision-making processes within the major standards-setting institutions? This is achieved primarily through working groups, and it is in these working groups where the policymaking process formally occurs.

At ISOC, the formal process for a proposed specification to become an official Internet standard is as follows. The process includes evolving through a set of maturity levels known as the "standards-track"; from at least 6 months as a Proposed Standard, to at least 4 months as a Draft Standard, and finally to an Internet Standard.[27]

Central to this process of graduating from one level to the next is one particular group within ISOC known as the Internet Engineering Steering Group (IESG). Any "standards action," which includes entering a specification into, advancing it within, or removing it from, the standards-track, must be approved by the IESG. If a specification has remained at the same maturity level for 24 months, the IESG can review the viability of the standardization effort and approve either the termination or continuation of the development effort.[28]

The standards-track process begins when a technical specification is posted in the "Internet-Drafts" directory where it will undergo two weeks of community review. At that point, a recommendation is made by the appropriate IETF Working Group, and it is up to the working-group chairperson to decide if a rough consensus has been achieved. Based on that recommendation, the IESG determines if the proposed specification satisfies the requirements for graduating to the next level in the process.[29]

As the specification nears final approval, the IESG issues a Last-Call Notification for final comments and community review that typically lasts no shorter than two weeks. This Last-Call Notification is sent via e-mail to the IETF Announce mailing list to permit a final review by "the general Internet community." Comments are accepted from anyone. The IESG then makes its final determination whether or not the specification will become a formal Internet Standard. Interestingly, the IESG is not bound by the action recommended when the specification was submitted.[30]

Upon approval, the standard is officially formalized via publication in the ISOC newsletter. A notification is sent to the RFC editor to publish the specification as an RFC. The specification is then removed from the "Internet Drafts" directory.[31]

Because technical standards and protocols are essentially a form of policy, in the context previously discussed, this process of creating such standards and protocols is the Internet's version of a policymaking process. Many of the same elements can be found in ISOC's standards-track as are recognized in the Political Science literature. A focusing event in the form of a technological innovation, seized upon by a policy entrepreneur, initiates the movement. A proposal is submitted to the formal institution responsible for creating such policies, at which point it is designated to an appropriate working group, or committee. After repeated efforts at community review and requests for comments, the committee chairperson decides if the proposal ought to go before the rest of the institutional body. After another period of community review by the larger body, the IESG makes its final determination whether to legitimize the policy. If so, a formalization process occurs as the proposal becomes an official standard. Finally, implementation of the standard is expected to occur—mainly in the private sector—and a period of continuous evaluation commences.[32]

A similar pattern of policymaking emerges at the W3C. A "Recommendation" proceeds through 5 maturity levels: Working Draft, Last-Call Working Draft, Candidate Recommendation, Proposed Recommendation, and finally an official W3C Recommendation.[33] Any such W3C "Process Documents" undergo a similar process of building rough consensus in working groups, with an advisory board acting as the sponsoring working group by putting out a request for comments, then calling for an advisory committee review lasting at least 4 weeks, and finally, if a consensus has been reached, announcing the W3C decision.[34]

The W3C, like ISOC, also leaves implementation of its Recommendations up to private-sector manufacturers. Additionally, many of its standards define levels of conformance, which the developers must follow if they wish to label their product W3C-compliant.[35]

Policies are thus being made at the Protocol layer by the working groups within these institutions using the rough consensus principle. However, what are the similarities and differences between this rough consensus model and other forms of policymaking? What lessons about governance and policymaking can be drawn?

To a large extent, the rough consensus model reflects elements of what Charles Lindblom classically referred to as "muddling through" the policy process. The ongoing processes that the major consortium groups use to design and formalize technical protocols, as previously described, clearly foster a method of Successive Limited Comparisons, in contrast to a Rational Comprehensive approach, insofar as they encourage adjustments at the margin. This is often brought on by new technological innovations. The example of IPv6 supports this notion as its design signaled an incremental change from its technical predecessor, IPv4, rather than a complete shift in values. Lindblom's assessment of muddling through is applicable to how the working groups within ISOC, the IETF, and the W3C function: They do not "find general formulations of objectives very helpful and in fact make specific marginal or incremental comparisons. . . . The only values that are relevant to [their] choice are these increments by which the two policies differ."[36]

According to the model of Successive LimitedComparisons, in fact, the test of a "good" policy is not one that can be shown as the most appropriate means to a desired end, as the Rational Comprehensive model would suggest, but rather is one where various analysts find themselves in direct agreement.[37] This is practically the very definition of rough consensus.

However, the rough consensus principle is not, by any means, a perfect reflection of muddling through. Lindblom downplays the role of values in decision-making, and asserts that administrators "need not try to analyze any values except the values by which alternative policies differ."[38] This "value problem" stands in direct contrast to the idea of the major consortium groups constituting an epistemic community. As described in the previous chapter, policies are being either influenced (at the Infrastructure layer) or outright designed and adopted (at the Protocol layer) by an international community of scientists and engineers who hold a core central value—optimum technical efficiency—as their common episteme. Whichever issues at these two layers are analyzed—whether U.S. telecommunications policy or the deployment of the TCP/IP protocol—this shared value is clearly the driver of policy design.

While the working groups within ISOC, the IETF, and W3C all operate under a system of Successive Limited Comparisons, and have developed open processes that foster such a system, they are only muddling through to a certain extent. The larger epistemic community of which they are a part remains very value-centric.

Two important things stand out about the rough consensus model for policymaking. First, these standards-setting institutions governing the Internet's protocols typically have open membership and open and transparent processes, constituting an epistemic community, and thus are fundamentally different types of institutions than, for example, the U.S. Congress. They are international in scope, virtually anyone can design, submit, and vote on policy proposals, there is no veto on policy outcomes, the role of technical expertise is paramount, and engineering efficiency is the prevailing ideology. Second, technological innovation plays a major role in driving, what is referred to in Institutional Rational Choice theory as, new "action arenas."[39] It is a policy catalyst that goes beyond problem definition to redefining new participants, outcome possibilities, and action-outcome linkages.

As a result, the utility of the rough consensus model to policymaking in other nontechnical arenas is greatly limited. As the governance of the Internet's protocols have demonstrated, rough consensus works best when institutional membership is fluid and when interests, though not the guiding larger ideologies, are frequently realigned—in this case, on the basis of rapid change in the technological landscape. The technical decisions over protocols are a major cause of cascaded effects on how software developers constrain or enable end-users' interaction with the Internet. This is what we shall explore in the next chapter.

NOTES

1. TCP/IP overcomes differences among networks by shifting the responsibility for technical reliability away from the network itself and instead toward the host devices. See Barry M. Leiner, Vinton Cerf, David D. Clark, Robert E. Kahn, Leonard Kleinrock, Daniel C. Lynch, Jon Postel, Larry G. Roberts, and Stephen Wolff, *A Brief History of the Internet* (2003). Retrieved on September 15, 2012 from http://www.internetsociety.org/internet/internet-51/history-internet/brief-history-internet.

2. Lessig, *Code*, 32.

3. Denardis 5.

4. *ISOC Official Website*. Retrieved on March 3, 2009 from http://www.isoc.org/isoc/.

5. *ISOC—List of Organisation Members*. Retrieved on March 24, 2009 from http://www.isoc.org/orgs/members.php.

6. *ISOC—List of Organisation Members*. Retrieved on March 24, 2009 from http://www.isoc.org/orgs/members.php.

7. *ISOC—List of Organisation Members*. Retrieved on March 24, 2009 from http://www.isoc.org/orgs/members.php.

8. *IETF Official Website*. Retrieved on March 3, 2009 from http://www.ietf.org/overview.html.

9. *Carolyn Duffy Marsan*, "Q&A: Security Top Concern for New IETF Chair," *Network World* July 26, 2007. Retrieved on March 3, 2009 from http://www.networkworld.com/news/2007/073007-ietf-qa.html.

10. *IETF—Past Meetings*. Retrieved on March 3, 2009 from http://www.ietf.org/meetings/past.meetings.html.

11. *W3C Official Website*. Retrieved on March 3, 2009 from http://www.w3.org/Consortium/.

12. *W3C—List of W3C Members*. Retrieved on March 3, 2009 from http://www.w3.org/Consortium/Member/List.

13. *W3C—About W3C Membership*. Retrieved on March 3, 2009 from http://www.w3.org/Consortium/membership.

14. *W3C Official Website*. Retrieved on March 3, 2009 from http://www.w3.org/Consortium/.

15. *W3C—Membership Fee Calculator*. Retrieved on March 25, 2009 from http://www.w3.org/Consortium/fees.

16. Peer Zumbansen, "Rough Consensus and Running Code: A Theory of Transnational Law Making," Paper presented at the Annual Meeting of the The Law and Society Association, Berlin, Germany, July 25, 2007. Retrieved on March 4, 2009 from http://www.allacademic.com/meta/p175687_index.html.

17. David D. Clark, "A Cloudy Crystal Ball: Visions of the Future," Plenary presentation at the 24th meeting of the Internet Engineering Task Force, Cambridge, MA, July 13–17, 1992. Slides from this presentation retrieved on March 4, 2009 from http://ietf20.isoc.org/videos/future_ietf_92.pdf.

18. *IETF Working Group Guidelines and Procedures*. Retrieved on March 4, 2009 from http://tools.ietf.org/html/rfc2418.

19. Andrew L. Russell, "'Rough Consensus and Running Code' and the Internet-OSI Standards War," *Annals of the History of Computing, IEEE* 28.3 (July–September 2006): 48–61.

20. Russell, 48–61.

21. Lessig, *Code*, 4.

22. Russell, 48–61.

23. William Drake, "The Internet Religious War," *Telecommunications Policy* 17.9 (December 1993): 643.

24. M. A. Padlipsky, *Elements of Networking Style* (Englewood Cliffs, NJ: Prentice-Hall, 1985), 104.

25. Einar Stefferud, quoted in M. T. Rose, "Comments on 'Opinion: OSI Is (Still) a Good Idea'," *ConneXions* 6.8 (1992): 20–21.

26. Denardis, 4.

27. *IETF—The Internet Standards Process—Revision 3*, Network Working Group, "Request for Comments 2026," (October 1996). Retrieved on March 26, 2009 from http://www.ietf.org/rfc/rfc2026.txt.

28. *IETF—The Internet Standards Process—Revision 3*.

29. *IETF—The Internet Standards Process—Revision 3.*

30. *IETF—The Internet Standards Process—Revision 3.*

31. *IETF—The Internet Standards Process—Revision 3.*

32. This reflects Lowi's heuristic stages of public policymaking: problem definition, formulation, adoption, implementation, and evaluation. See Theodore J. Lowi, "Four Systems of Policy, Politics, and Choice," *Public Administration Review* 11 (1972): 298–310.

33. *W3C Process Document* (October 14, 2005). Retrieved on April 10, 2009 from http://www.w3.org/2005/10/Process-20051014/.

34. *W3C Process Document—12 Process Evolution.* Retrieved on April 10, 2009 from http://www.w3.org/2005/10/Process-20051014/processdoc.html#GAProcess.

35. *W3C Process Document* (October 14, 2005).

36. Charles E. Lindblom, "The Science of 'Muddling Through'," *Public Administration Review* 19.2 (Spring 1959): 79–88.

37. Lindblom, 81.

38. Lindblom, 84.

39. The concept of action arenas is common throughout the Institutional Analysis and Development (IAD) framework and is meant to include both "action situations" and the actors in those situations. It includes seven clusters of variables: participants, positions, outcomes, action-outcome linkages, the control that participants exercise, information, and the costs and benefits assigned to outcomes. See Elinor Ostrom, "Institutional Rational Choice: An Assessment of the Institutional Analysis and Development Framework," *Theories of the Policy Process*, ed. Paul A. Sabatier (Boulder, CO: Westview Press, 1999), 35–71.

Chapter 5

Who Governs the Applications?

Cyberspace is a virtual environment that only exists through software. As such, who creates that software is, fundamentally, engaged in a type of policymaking. Software applications are what enable human beings to interact with the digital network, and therefore determining who is developing that software, and what political values the end-products come to embody, is vitally important.

In this chapter, we will examine the ways in which the code underlying both desktop and web applications is a form of policy itself. These software applications enable and constrain the actions of every Internet user on a technical basis, and thus we will demonstrate how code constitutes a unique type of policy, one in which the environment itself is designed to deny the user even a *capability* to act in defiance. While any individual with the requisite computer programming skills can reasonably be said to take part in the governance equation insofar as they set the rules for their own private cyberspaces, we will analyze certain Internet usability metrics and argue that, while individual nonaffiliated programmers hold a significant, often disruptive influence, that has yet to translate into governing authority when looking at the Web in aggregate. As a result, we will argue that a relatively small handful of the most well-capitalized private commercial software firms govern the Internet's applications the most—again, based on usability metrics. Ultimately, we will explain how Lessig's "code is law" argument best explains how code constrains and enables Internet behavior, only, we will argue, that the code written by private commercial firms often indicates an implicit recognition of the sovereign authority that traditional governmental institutions retain over them.

WHY ARE THE APPLICATIONS IMPORTANT?

Software applications are the tools which allow people to make use of the Internet. Once an infrastructure connects disparate devices, and once technical standards and protocols facilitate communication between those devices, individual people then require an interface, or method of interacting, with the digital network. Software applications perform this function.

In Yochai Benkler's original layer model, the "code" layer encompassed both the technical protocols as well as software applications. While this may have been appropriate from his more narrow scope of media regulation, when the model is applied to Internet governance more generally, it leaves too much unanswered. From a policymaking perspective, there is a tremendous difference between how policies are made when it comes to technical protocols versus software applications; as well as a crucial distinction between who is making those decisions. Protocols and applications, in the context being discussed, are entities of fundamentally different types, and thus Benkler's "code" umbrella used to encompass them both needs to be revised in order to account for those differences.

The software applications at the heart of the debate refer to Internet- and Web-specific software applications. For example, web browsers like Microsoft's Internet Explorer, Apple's Safari, Google's Chrome, and the Mozilla Foundation's Firefox are software programs that play a major role in determining how people interact with the Internet. How those browsers are programmed ultimately translates into enabling certain forms of cyberspatial behavior while simultaneously limiting others. Therefore, power over the software is the power over people's actions in cyberspace.

This holds true not only for web browsers but also for other types of software people use to interact online including operating systems like Microsoft Windows, stand-alone applications like Adobe Acrobat, and server-side programs like the Apache Web Server. Corresponding programming languages behind these applications, such as C++, Java, PHP, and others, are similarly tools whose design either constrains or enables different types of behavior.

The Applications layer encompasses not only proprietary pieces of software that are purchased and installed on a device, but also those that are classified as Web Services. As people's digital activities have migrated increasingly from the desktop to the Web, the applications and services delivered by websites have become equally important as the programs installed locally on one's computing device. For example, the shopping cart technology used by Amazon.com, the blogging services provided by Blogger or WordPress, the web-based e-mail services offered by Yahoo or Hotmail, the content-sharing features of Facebook or YouTube, or the participatory editing features of Wikipedia, are all characteristically software applications, written in code and programmed by people, that also either constrain or enable different types of behaviors for their users. This is, undeniably, a form of governance.

Ultimately, software applications are important because, whereas the infrastructure and protocols dictate how data is transported over the network, it is the software which sets the rules for how people will participate in online activities.

WHO GOVERNS THE APPLICATIONS?

The software applications and web services that allow people to interact with the Internet are governed directly by the computer programmers who write their underlying code. While this can theoretically encompass virtually everyone, in practice, based on usability metrics, it translates into private commercial software firms being the primary agents of governance and policymaking at the Applications layer.

Determining governance at the Applications layer is problematic. Based on the assumption that the code underlying software is the central mechanism that constrains and enables different types of behavior, it follows that any individual, private entity, or public agency who writes code can, therefore, be said to govern, or at least to take part in the governing process.

However, this notion, while true in theory, doesn't help us understand the complex power arrangements in the software development community. Because this community, like the Internet itself, is highly decentralized, attempts by national governments to regulate it are inherently limited. To reference an established First Amendment principle, there is no "prior restraint" when it comes to government intervention in the production of code. At most, there is only attempted regulation of the finished software product.

However, while any programmer can certainly be said to govern the specific private cyberspaces that their code creates, when it comes to analyzing the Web in aggregate, more definitive and demonstrable patterns of governance do emerge. Some actors, indeed, can be accurately said to govern more than others.

So in this decentralized scenario where virtually anyone with programming knowledge can write code and create software, potentially distributed and used by millions of people throughout the world, what methodology can we employ to come to a more concrete understanding of who, exactly, is governing at the Applications layer?

In actual practice, a relatively few private commercial firms hold considerable authority in setting the rules for behavior. Evidence of this can be traced back to our original definition of what it means to govern: to have the ability to both constrain and enable different types of behavior with intentional effects.

Thus, we can determine who exactly has governing power by examining how code grants authority to programmers to shape cyber environments and, subsequently, by examining the two primary constraints that act over them.

Fundamentally, code grants computer programmers the ability to shape the virtual environments in which the average user interacts. This is Lawrence Lessig's "code is law" principle which will be explained more thoroughly in the following section. For example, the programmers behind the social-networking site Facebook have made the decision to allow anybody to create groups, on any topic, without requiring any prior editorial approval. Conversely, Facebook has also made the decision not to permit people to stream audio clips from their profiles, as some of its rivals do, for fear of copyright concerns. In either case, these decisions, whether to enable or constrain people's behavior, are accomplished by programmers primarily through code.

Lessig is absolutely correct in arguing that, as opposed to real-space, these policies are not simply a matter of establishing a law and enforcing it by creating penalties for people who break the law; rather, it is an altogether different type of law, one in which the environment itself is designed to deny the user even a capability to act in defiance.

Not only is it through code that programmers are empowered with such significant governing power insofar as they shape their virtual environments, but code also dictates the actions of those very same programmers and places constraint on them. Using the above example, while it's true that Facebook's programmers can code the policies of their own private virtual environment as it affects the end-user, it is also true that those programmers must still adhere to sets of rules that have previously been established. In other words, someone else has authority over them, and that too is an authority derived from code.

The first constraint that acts over programmers is language. While it may be something of blasphemy within certain circles of the programming community to make this assertion, it nevertheless holds true. A C++ or .NET programmer may believe that they can write code to do whatever they want on a technical level, however that is only true within the confines of what Microsoft decided C++ as a language would allow. The designs of all computer programming languages are the result of explicit decision-making processes, often by formal institutions, and those decisions ultimately have consequences on the resulting decisions of those who implement them. In other words, *the capabilities and limitations of programming languages act as inherent checks on the behavior of programmers.*

This principle also raises the second constraint on programmers—the computing platform. Many programmers might reluctantly concede to the above assertion, but they will then undoubtedly point to alternatives that are not controlled by private commercial firms. For example, a web programmer might argue that if he contributes to the development of a nonproprietary open-source language like PHP, Perl, or Python, then language becomes less of a restriction because he can have a hand in shaping it. However, the programming language is not the only constraint on the programmer. Even if one created an entire programming language from scratch, the behavior of the programmer would still be determined by the platform on which the resulting software would be used. The code behind such platforms, whether an operating system like Microsoft Windows, non-OS-dependent platforms like Sun's Java, or various "application programming interfaces" like the Google API, also either constrains or enables the behavior of programmers. In other words, *a programmer's code, no matter how independent, must still be written within the confines of rules established by the platform if it wants to achieve a reasonable level of operability.*

The evidence for how private commercial software firms are the primary agents of governance at the Applications layer lies in usability. While it's true that, in theory, anyone can create a piece of software and set the rules for behavior, in actual practice the overwhelming majority of Internet users only use a relatively small number of software applications and web services—and of these, the vast majority tend to be operated by private commercial firms.

To measure usability, it is first necessary to conceptually differentiate between the two types of applications that are our focus—desktop applications vs. web applications. Both are critical in understanding how code governs.

Desktop applications refer to any software that can be installed on a single computer and used to perform specific tasks. Some desktop applications can also be used by multiple users in a networked environment, but, by definition, it is the software that is installed and then resides on one's computer, and follows the traditional business model of consumers purchasing a disk or CD in

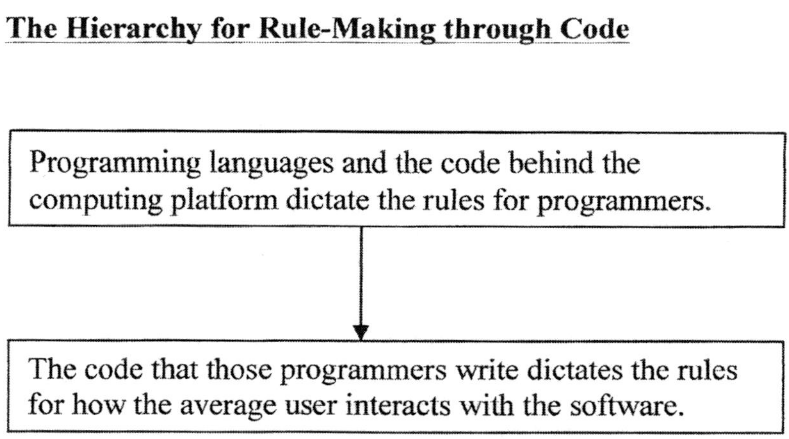

Figure 5.1 The Hierarchy for Rule-Making Through Code

a shrink-wrapped box and "owning" their copy.[1] Common examples include the Microsoft Office Suite.

In contrast, web applications often form a desirable alternative to desktop applications for reasons of portability. Web applications are usually based on a client-server architecture and use a web browser as the client interface. The software exists, not on a user's hard drive, but in cyberspace. It is a software-as-service model that functions either through the client's web browser or through "cloud computing," where examples like shopping cart applications, online video archives, remote file storage services, photo managers, calendars, accounting programs, and more have become ubiquitous.[2]

It is next necessary to establish a set of metrics by which we can accurately assess, using quantitative data, which specific desktop and web applications are used the most—and also assess qualitative factors, such as the weighted impact of that software.

The most important types of desktop applications, as it relates to affecting people's interaction with the Internet, are the operating system and the web browser. The reason for this is that, returning to the two main constraints that act on programmers, operating system software is the primary computing platform on which web applications consequently run, and, additionally, web browsers are functionally the most common interface used for interacting with the Web. An analysis of each indicates that private commercial firms clearly dominate their respective markets, despite the existence of viable non-proprietary alternatives:

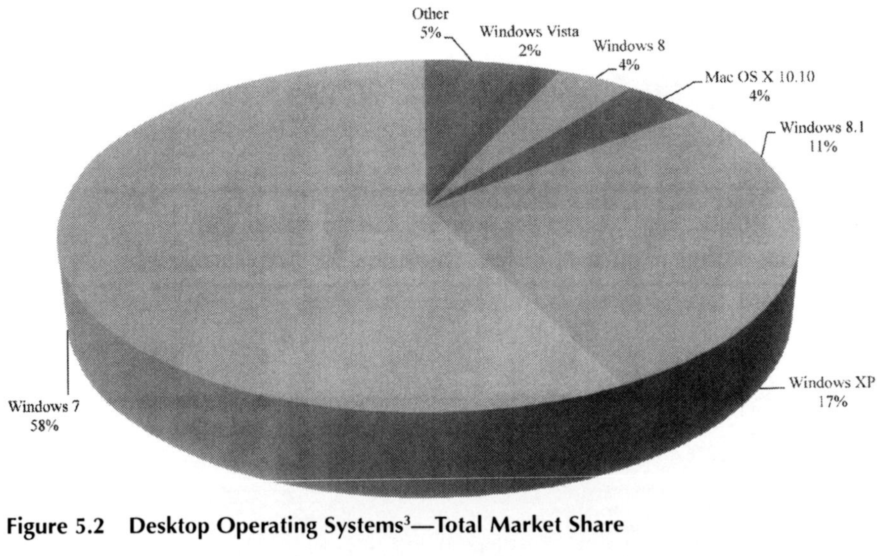

Figure 5.2 Desktop Operating Systems[3]—Total Market Share

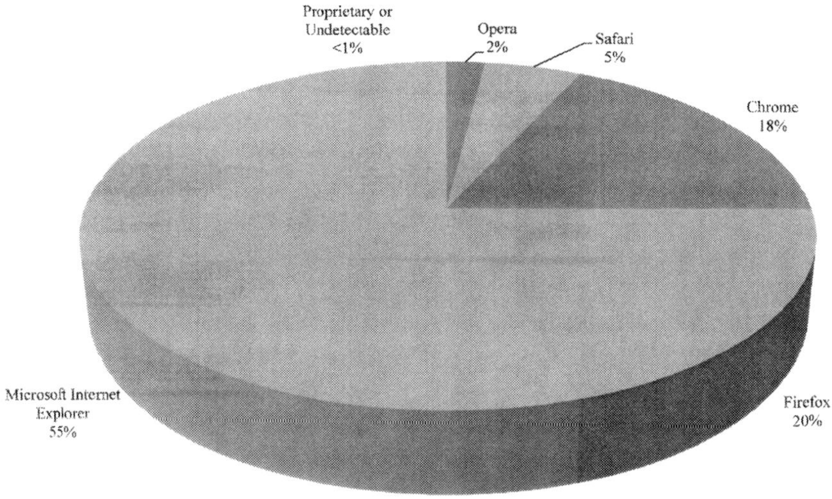

Figure 5.3 Desktop Web Browsers—Total Market Share

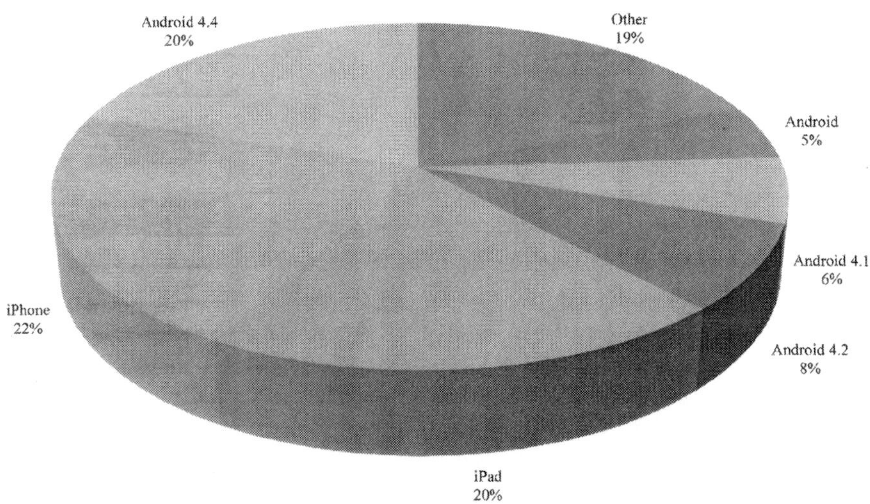

Figure 5.4 Mobile/Tablet Operating Systems[4]—Total Market Share

Clearly, when it comes to operating systems, Microsoft, as the owner of Windows, has a tremendous governing power over the space with over a 90 percent market share, and all but Linux on this list (included in the "Other" category) are controlled by either Microsoft or Apple. Similarly, as far as web browser software is concerned, Microsoft again has a clear dominance in the space in terms of usability with its Internet Explorer software; and once again, private commercial firms in general clearly dominate the list of most-used browser software; this despite the relative popularity of Firefox in recent years, notable because it is controlled by a nonprofit organization, the Mozilla Foundation.

The pattern repeats itself when examining the mobile/tablet space where private commercial firms—in this case, Google and Apple—once again dominate.

What this data demonstrates is that, insofar as they control the computing platform upon which web applications are designed to operate, and upon which the user most typically uses to interact with the Web, Microsoft and other private commercial firms have a substantial power to set the terms by which computer programmers must subsequently adhere to when creating their code.

Furthermore, as to the constraint of languages, the TIOBE Programming Community Index ranks the most popular programming languages used by software engineers as follows[5]:

Table 5.1 Most Popular Programming Languages

1. Java	6. JavaScript
2. C	7. PHP
3. C++	8. Python
4. Objective-C	9. Visual Basic
5. C#	10. Visual Basic .NET

It is a familiar pattern once again. Just as private commercial firms have demonstrable governing authority via their control over the computing platform, so do they also have governing authority when it comes to the constraint of programming languages. Four of the top five languages were created and are maintained by private commercial firms, with Microsoft controlling C, C++, and C#, and Apple controlling Objective-C. The sole exception at the top of the list, Java, was also created by a private commercial firm, Sun Microsystems, although it is released as open source. This is clear evidence that, because languages can both constrain and enable different types of behavior among programmers, private commercial firms have substantial governing power insofar as they maintain control over those languages.

Web applications paint something of a different picture. Once a user has booted their operating system and opened their web browser, the web applications they then interact with are exponentially greater in number and far more diverse in type than are the different desktop applications available to them.

Such diversity is still measurable, however. A set of metrics designed to gauge the quantitative aspects of usability in cyberspace must include the number of page hits measuring direct web traffic and the number of unique users of a web application.

The most essential type of web application is the search engine—the main portal, or jumping-off point, for cyberspatial activities.

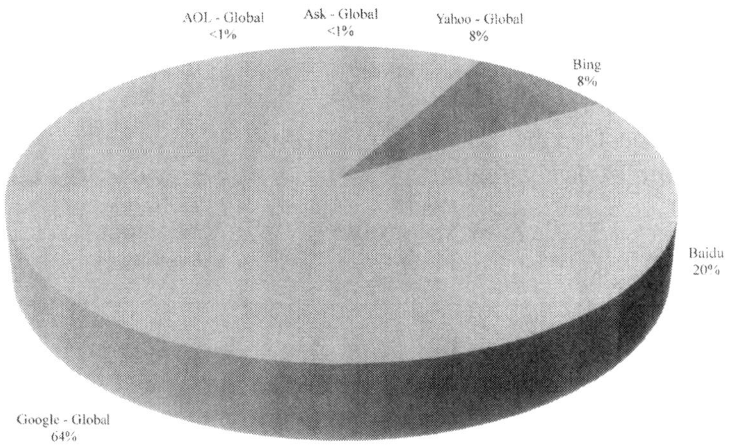

Figure 5.5 Desktop Search Engines[6]—Total Market Share

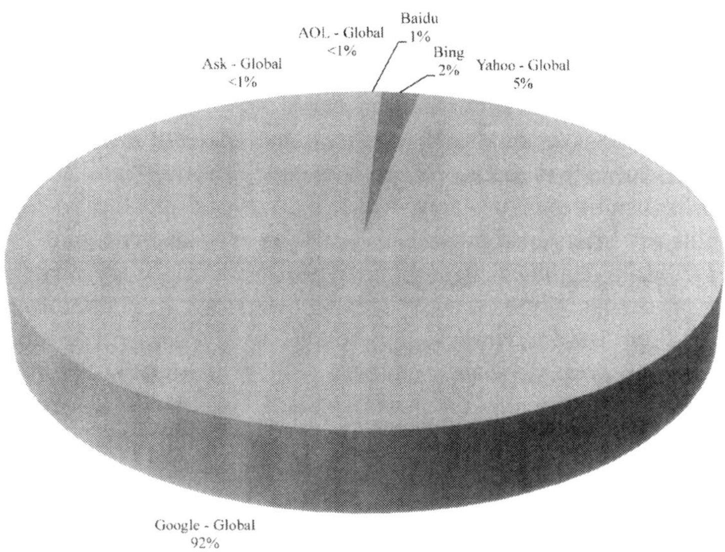

Figure 5.6 Mobile/Tablet Search Engines[7]—Total Market Share

Google is the premier actor in the search engine space, and again, the list of top applications is completely dominated by private commercial firms.

When it comes to websites specifically, a similar usability pattern emerges. For example, an examination of the top 100 websites based on Alexa rankings reveals that, of the top 100 most visited websites, over 90 percent are operated by private commercial firms. Using only the quantitative metric of number of page hits, the most-used websites in cyberspace are the following:[8]

Table 5.2 Most Used Websites (by number of page hits)

1. Google	10. QQ.com	18. Google Japan
2. Facebook	11. Google India	19. EBay
3. YouTube	12. Live.com	20. Blogspot
4. Yahoo	13. LinkedIn	21. T.co
5. Baidu	14. Sina	22. Google Germany
6. Wikipedia	15. Weibo	23. Yandex.ru
7. Amazon	16. Yahoo Japan	24. Hao123
8. Twitter	17. TMall	25. Bing
9. Taobao		

However, the proportionate leap in the impact from one of these sites to the next is not always of equal dimension—thus warranting a qualitative analysis to be performed. According to Nielsen/NetRatings, while American Internet users visit an average of 1,669 web pages each month, those pages reside on only 72 different domains.[9] What this means is that *a relatively few number of sites have a far greater impact and reach than others in terms of usage* because, while users may visit an average of 1,669 web pages per month, many of those different pages are actually owned or operated by the same parent organizations.

For example, a look at the top 10 parent organizations of the most popular websites reveals that they are all private commercial firms:[10]

Table 5.3 Top Parent Organizations of the Most Popular Websites

Parent Company	*Unique Audience (000)*	*Reach (%)*
Google	116,770	75.32
Microsoft	106,170	68.48
Yahoo	98,658	63.64
AOL	65,939	42.53
News Corp. Online	60,504	39.03
Facebook	51,370	33.13
InterActive Corp.	46,775	30.17
EBay	45,993	29.67
Amazon	42,588	27.47
Apple Computer	40,133	25.89

In the above data, "Reach" is the "percentage of all active unique visitors who visited the site or used the application during the month. Active is defined as anyone who used an Internet-enabled computer during the month."[11]

These parent organizations have tremendous power, through code, to both enable and constrain different types of people's behavior in cyberspace; and because people use these organizations' websites the most—by a substantial amount in some cases—their governing authority is very real, despite the fact that any programmer can independently write code and be said to govern their own private sites. By looking at usability, some entities clearly govern far more than others.

Furthermore, *when we look at these metrics that show how people are actually using the Internet, it is clear that they are using it almost completely in a way that gives private commercial firms authority over their behavior.* Based on the Internet's decentralized architecture, as well as the decentralization of computer programmers writing code, this need not necessarily have to be the case, but, based on usage, it is nevertheless.

This conclusion confirming the dominance of private commercial firms in web applications, however, is complicated when analyzed through the lens of the constraint of programming languages. Unlike with desktop applications, where the most commonly used C-family languages are controlled by private commercial firms, such is not the case with web applications. Reviewing the aforementioned list, the next top-10 languages are JavaScript, PHP, and Python. These (not to mention standards-based languages like HTML and XML) are all nonproprietary. No single company or group of companies controls web-based languages or singularly guides their development. Rather, these languages are continually developed by volunteer programmers and software engineers, often without any institutional affiliation or compensation.[12]

Throughout the history of computing, programming languages from Fortran, Basic, and Cobol to C++, Java, and PHP have enabled, and, in fact, encouraged, a decentralized approach to the generation of new software. For example, one does not need to be a Microsoft employee in order to write code in C++. In fact, Microsoft created the C++ language with the express intent of empowering outsiders to create software on their own, and to do so with no lingering association back to the company. As a result, computer programmers do not necessarily write their code as part of any formal hierarchy or within any institution that can be accurately said to have oversight capabilities—other than, perhaps, the organizations which employ some of them, however, even those organizations, whether public or private, may not have any direct associations with each other. This decentralization and empowerment of individual, nonaffiliated programmers eventually led to the open-source movement in computing, whereby programmers are encouraged to openly share their source code for the sake of collaboration and peer production of software—forming a viable alternative to the closed, proprietary model administered by most private commercial firms.[13]

This nebulous open-source community is extremely significant and can stake claim to numerous highly successful software applications.[14] For instance, it is responsible for highly used web applications such as the Apache Web Server, which hosts approximately 52 percent of all websites on the Internet. By comparison, Microsoft's IIS, the next most-used web server, can claim approximately 33 percent.[15]

It is in this web application context where many voices in the programming community firmly assert that private commercial firms do not have authority over their actions; and to a certain extent, they're correct. When considering how many of the most popular web programming languages are nonproprietary, and considering how similarly open-source web software like the Apache Web Server is actually used to host more websites than its commercial rivals, it would be intellectually dishonest to simply write off these phenomena as an inconvenient sideshow. In fact, such nonproprietary languages and open-source web applications form the basis of what ought to be construed as a very significant component in the governance formula—or at the very least a great mitigator on the powers that private commercial software firms hold.

To clarify this point, there are specific programmers who have had serious disruptive influences throughout the Internet's history—programmers such as Shawn Fanning (creator of the original Napster program, which introduced millions to illegal downloading of copyrighted music files), Phil Zimmerman (creator of the PGP application, which enhances people's digital privacy but makes law enforcement and investigation far more difficult), and Linus Torvalds (creator of Linux, the open-source operating system). The characteristics that this group of individual nonaffiliated programmers share in common is that each worked independently, outside of any public or private institution, utilizing nothing more than their programming skills and a personal computer to develop software for other people to use that was then disseminated online to the public for free. This proved disruptive in that the software they created and released either (1) facilitated the outright subversion of U.S. law (Napster), (2) was legal but ran counter to local, state, and federal objectives (PGP working against certain national security objectives), or (3) was legal but ran counter to the financial interests of private commercial firms (Linux and open-source programs undermining firms like Microsoft and Apple).

In response to the effects wrought by these individual nonaffiliated programmers, firms such as Google, Facebook, Microsoft, Apple, and others have adopted new policies and implemented new technologies that grant them more centralized control. For example, YouTube has implemented specialized software to filter and remove clips that potentially violate copyright law. Meanwhile, other commercial institutions such as the Recording Industry Association of America (RIAA) and Motion Picture Association of America (MPAA) have litigated cases and lobbied in Washington D.C. to pass federal laws that enhance protections of their commercial interests, while still others have sought to limit their own legal liability, as was the case with ISPs like AOL in the policy process leading to the Digital Millennium Copyright Act of 1998.[16]

As a result, in their various responses to the disruptive power of individual nonaffiliated programmers—whether those responses take the form of preventative, reactionary, or coordinative actions—private commercial firms prove to be the primary governing authority. Ultimately, individual nonaffiliated programmers have significant, and often disruptive, *influence*, even in creating Internet policies (in their capacity to write the code underlying much of the software). However, that has failed to translate into *governing authority*.

When looking at usability from the perspective of the average user's Internet experience, it remains true that, despite the widespread prevalence of nonproprietary programming languages and open-source web applications, their cyberspatial behavior is still primarily determined by the code written by private commercial firms. This is not to say that private commercial firms govern all of the activities of all users on the front-end, or even the development of all web technologies on the back-end, for that is certainly not the case. Numerous alternatives do indeed exist and some are even highly successful in their deployment rates. But what the data does undeniably indicate is that the majority of what the average Internet user does, and is capable of doing, in the websites they visit, the applications they encounter, and the services they use, still falls under the authority of private commercial software firms. These firms do, in fact, govern more.

HOW ARE POLICIES BEING MADE AT THE APPLICATIONS LAYER?

Policies are being made at the Applications layer primarily by private commercial firms through the writing of code. The fundamental question is what political values that code represents. At the center of the debate over governance at the Applications layer is Lawrence Lessig's argument that "code is law."

To clarify, the focus of governance at the Applications layer is how cyberspace, in particular, is governed; not the Internet as a whole. Cyberspace refers specifically to the virtual environments where people's online behavior takes place. It does not refer to the infrastructure or hardware elements that make up the Internet. Rather, cyberspace, in the context being addressed, is its software component.

As such, this software component is created exclusively by code. Computer programming languages are used to generate a particular software environment, essentially establishing the laws of nature—what is technically possible and what is not possible—within a particular realm. In real-space, governments may enact certain types of laws for its citizens to follow, however, other environmental laws—such as the laws of gravity and physics—remain in place, and governments are powerless to trump them.

However, the fundamental difference between real-space and cyberspace is that, in cyberspace, the so-called environmental laws can be programmed as well. In other words, gravity actually can be legislated away in cyberspace. Virtual environments, written entirely in code, are designed upon a far more malleable foundation than the physical laws of real-space. As Lawrence Lessig has described, "problems don't necessarily have to be solved; they can be programmed away. Rather than resolve conflicts between two people by making one of them change their behavior, it's possible to simply change the laws of nature to eliminate the conflict altogether."[17]

Because the environment of cyberspace is created through software written entirely in code, Lessig argues that code is the law of cyberspace. "In real space we recognize how laws regulate—through constitutions, statutes, and other legal codes. In cyberspace we must understand how code regulates—how the software and hardware that make cyberspace what is it *regulate* cyberspace as it is."[18]

The types of policies being created by code throughout cyberspace, constraining or enabling the capabilities of the environment itself, are illustrative examples of governing authority, according to our definition. Code is serving not only as a tool for policy implementation, but its underlying decision-making processes are where those policies actually originate.

With the "code is law" argument established, the question, then, is which political values code has come to represent. Policy objectives—or the intents of a particular software application—are, as in real-space, political decisions that embody different sets of values and priorities.

> This code presents the greatest threat to liberal or libertarian ideals, as well as their greatest promise. We can build, or architect, or code cyberspace to protect values that we believe are fundamental, or we can build, or architect, or code cyberspace to allow those values to disappear. There is no middle ground. There is no choice that does not include some type of *building*. Code is never found; it is only ever made, and only ever made by us.[19]

From an engineering perspective, the policies, or rules, of a particular piece of software are coded into its architecture. Because that architecture can either be programmed to make a virtual space more democratic or authoritarian, to be more openly participatory or editorially controlled, or to allow for anonymous behavior versus requiring the authentication of identities. Lessig refers to these decision-making outputs as varying "architectures of control"—and the extent to which control is coded into the architecture will directly affect people's behavior. Whether or not a specific virtual environment can be regulated, and how that regulation is designed and implemented, turns on the nature of its code.[20] Some architectures are more regulable than others. As a result, *in cyberspace, software design is a form of policymaking, and its architecture is its politics.*[21]

This idea that software code ought to be considered a meaningful form of policy is not, however, universally agreed upon. While Lessig's "code is law" argument has been the evocative center of the academic debate, other scholars like Tim Wu have been quick to highlight "when code isn't law."[22]

Wu takes issue with the "grand speculation" that has resulted in the years since Lessig's original argument was published, citing broadly interpretive claims ranging from how code will arise as a type of utopian sovereign, to how code may be used to negate basic freedoms, to how code shouldn't even be considered a legal novelty at all. Instead, Wu proposes to study the design of code as an aspect of interest group behavior. Doing so, he concludes, reveals the need to differentiate between two separate aspects of code's relationship to law—first, code's ability to act as a regulatory mechanism in substitute for traditional laws or other forms of regulation; second, code's ability to act as an antiregulatory mechanism, meaning "a tool to minimize the costs of law that certain groups will use to their advantage."[23]

To clarify, Wu's counterargument to Lessig is that code, rather than being a new type of law, is actually a mechanism for avoidance from the law. Using P2P file-sharing software as his primary example, he argues that code simply allows specific computer-savvy groups to take advantage of loopholes and legal ambiguities to afford them a tool for noncompliance. This is hardly the great paradigm shift that many of Lessig's disciples have suggested.

Furthermore, other counterarguments to the "code is law" principle have emerged as well. Jack Goldsmith and Tim Wu have together put forth a direct challenge to the assumption that the Internet liberates us from traditional legal norms imposed by governments and geographic borders. Territorial governments have re-asserted themselves in cyberspatial issues over the last decade and, Goldsmith and Wu argue, rather than code becoming law, or even resisting territorial law, code is now actually facilitating its enforcement.[24]

So which is it? Is code the law of cyberspace, or is it just a facilitator, or a conduit, of the traditional political power of national governments?

It is both. To understand how, let us re-iterate that, based on usability, it is private commercial software firms who are primarily creating the policies and designing the software architectures at the Applications layer. This sheds light into exactly how policies are being made, and it is a question that can, as a result, be re-phrased as follows: How are private commercial firms making decisions as to how to code their software?

Lessig is right that code is the law of cyberspace in that it unquestionably dictates the rules for how people can interact with it. However, Goldsmith and Wu are also right. Because private commercial firms are writing the code and designing the software architectures that people are overwhelmingly using in cyberspace, and because those private commercial firms adhere to the jurisdictional laws of territorial governments in which they are incorporated and in which they operate, they are unquestionably deciding what form their code will ultimately take based on the territorial laws which apply to them. To state it plainly, private commercial firms want to gain entry into markets and will modify their code to comply with local laws in order to do so. Lessig himself, in fact, writes that the aggregate transformation that cyberspace is experiencing toward more regulable architectures is not the product of governments—they are the product of user demand and deployed through commerce. He stipulates that these transformations are not the product of conspiracy, but rather they are "the consequence of changes made for purely pragmatic, commercial ends."[25]

For example, take the case of Google. The search engine that has an 84 percent market share earned in excess of $46 billion in total annual revenues in 2012,[26] handled approximately 1.2 trillion queries,[27] and is the single most-used website in cyberspace provides a model case study. Google's search algorithm itself favors certain values over others—it is based on a mathematical formula known as the PageRank algorithm which ranks search results based on the "authority" of websites and their external links. This algorithm, a form of code, has been criticized for reinforcing the authority of already-established entities, and stands in contrast to alternative architectures, such as that deployed by Reddit, which rely on voting-based ranking systems.[28]

Not only does Google's search algorithm demonstrate how code embodies certain political values, it also illustrates how commercial market forces and territorial laws substantially affect their decision-making processes. The company whose official motto is "Do No Evil" routinely engages in the censorship of search results if they view it in their commercial interests to do so. Google admits to censoring certain Nazi-related websites in Germany, child pornography sites in the United States, and a plethora of websites in China including those of various human rights groups and others covering politically sensitive subjects such as any relating to Taiwanese independence or the Tiananmen Square uprising of 1989.[29] They have made these decisions to remove or omit information from its services in order to comply with the laws of the different local and national governments in which they seek to operate in a commercial capacity.

Thus, code is, indeed, the law of cyberspace; but the code that people encounter most frequently is overwhelmingly written by private commercial firms whose actions indicate an explicit recognition of the sovereign authority that traditional governmental institutions have over them. Therefore, policies are being made at the Applications layer through decision-making processes that are primarily determined by commercial market forces, the technological capabilities of code, and territorial laws.

NOTES

1. Jeff Smith, "Desktop Applications Vs. Web Applications," *StreetDirectory.com*. Retrieved on Juy 29, 2009 from http://www.streetdirectory.com/travel_guide/114448/programming/desktop_applications_vs_web_applications.html.
2. Jason Tanz, "Desktop, R.I.P.," *Wired* (March 2007). Retrieved on April 1, 2007 from http://www.wired.com/wired/archive/15.04/wired40_rip.html.
3. Retrieved on April 19, 2015 from http://www.netmarketshare.com/operating-system-market-share.aspx?qprid=10&qpcustomd=0&qpcustomb=#.
4. Retrieved on April 19, 2015 from http://www.netmarketshare.com/operating-system-market-share.aspx?qprid=10&qpcustomb=&qpcustomd=1.
5. *TIOBE Programming Community Index for April 2015*. Retrieved on April 19, 2015 from http://www.tiobe.com/index.php/content/paperinfo/tpci/index.html.
6. Retrieved on April 19, 2015 http://www.netmarketshare.com/search-engine-market-share.aspx?qprid=4&qpcustomd=0#.
7. Retrieved on April 19, 2015 from http://www.netmarketshare.com/search-engine-market-share.aspx?qprid=4&qpcustomd=1.
8. "Top 500 Global Sites," *Alexa*. Retrieved on April 19, 2015 from http://www.alexa.com/topsites/global;0.
9. "Global Index Chart: Month of March 2009," *Nielsen Online*. Retrieved on May 5, 2009 from http://www.nielsen-online.com/press_fd.jsp?section=pr_netv&nav=3.
10. "United States Top 10 Parent Companies: Month of March 2009," *Nielsen Online*. Retrieved on May 5, 2009 from http://www.nielsen-online.com/reports.jsp?section=pub_reports&report=parent&period=monthly&panel_type=2.
11. E-mail from Michelle McGiboney, PR Manager, *Nielsen Online* (June 9, 2009).
12. See the epistemic community and international consortia groups described in Chapter 4.
13. Andrea Bonaccorsi and Cristina Rossi, "Comparing Motivations of Individual Programmers and Firms to Take Part in the Open Source Movement: From Community to Business," *Knowledge, Technology & Policy* 18.4 (2006): 40–64.
14. Sandeep Krishnamurthy, "Cave or Community?: An Empirical Examination of 100 Mature Open Source Projects," *First Monday* (2002). Retrieved on February 20, 2013 from http://ssrn.com/abstract=667402.

15. "January 2009 Web Server Survey," *NetCraft*. Retrieved on July 29, 2009 from http://news.netcraft.com/archives/2009/01/16/january_2009_web_server_survey.html.

16. *Public Law 105–304* (October 1998). Retrieved on February 20, 2013 from http://www.gpo.gov/fdsys/pkg/PLAW-105publ304/pdf/PLAW-105publ304.pdf.

17. Lessig, *Code*, 13. For example, Lessig raises a terrific case in Second Life where a woman named Martha grew and sold poisonous flowers. One of her "neighbors" named Dank owned a dog that ate one of the flower petals and died. The two engaged in a heated argument placing blame on each other. Dank didn't understand why Martha was selling poisonous flowers, but Martha couldn't understand why Dank had created a dog that suffered when dying (or was even susceptible to poison in the first place). Ultimately, their problem could best be resolved by re-engineering the code, or laws of nature, so that the flower petals would only be poisonous when in the possession of someone who purchased them as such, but when they were stolen or blown away, they would lose their poisonous qualities. See Lessig, *Code (version 2.0)*, 10–15.

18. Lessig, *Code*, 6.

19. Lessig, *Code*, 6.

20. Lessig, *Code*, 20.

21. Mitchell Kapor, "A Software Design Manifesto." Retrieved on May 29, 2009 from http://hci.stanford.edu/bds/1-kapor.html.

22. Tim Wu, "When Code Isn't Law," *Virginia Law Review* 89.4 (June 2003): 679–751.

23. Wu, "When Code Isn't Law," 682.

24. Goldsmith and Wu.

25. Lessig, *Code (version 2.0)*, 38.

26. "Financial Tables," *Google Investor Relations*. Retrieved on February 20, 2013 from http://investor.google.com/financial/tables.html.

27. *Google Zeitgeist 2012*. Retrieved on February 20, 2013 from http://www.google.com/zeitgeist/2012/#the-world.

28. Mike Thelwall and Liwen Vaughan, "New Versions of PageRank Employing Alternative Web Document Models," *ASLIB Proceedings* 56.1 (Emerald Group Publishing Limited, 2004). Retrieved on February 20, 2013 from http://www.emeraldinsight.com/journals.htm?articleid=864007&show=abstract.

29. Jeffrey Rosen, "Google's Gatekeepers," *New York Times* November 28, 2008. Retrieved on July 29, 2009 from http://www.nytimes.com/2008/11/30/magazine/30google-t.html?_r=1&partner=rss&emc=rss&pagewanted=all.

Chapter 6

Who Governs the Content?

When most people think of Internet issues their focus is on cyberspatial content. Whether a particular issue arises over censorship, privacy rights, photos distributed on social-networking sites, etc., these are all matters that center on the actual material that end-users see, read, listen to, download, watch, and interact with while online. As a result, the Content layer is the most highly visible, controversial, and politicized of all the four conceptual layers.

In this chapter, we will perform a detailed analysis of three Internet issue areas that have been highly prominent on legislators' agendas since the 1990s—the regulation of pornographic material online, the regulation of spam, and the regulation of file-sharing. What they will demonstrate is that governmental policies have often proven to be effective in enabling certain types of Internet content (e.g., Section 230 of the Communications Decency Act), while at other times governmental policies have proven to be inherently limited by the Internet's global scope (i.e.—court rulings on P2P file-sharing) or limited by the technical design of protocols (i.e.—SMTP and the CAN-SPAM Act). Thus, we will argue that while national governments certainly have governing authority over Internet content to an extent, ISPs and private website operators (through their TOS Agreements) also have demonstrated their authority to make policies that directly constrain or enable behavior with intentional effects, particularly in the transnational context. Ultimately, we will assert that adopting an Issue Network framework is most helpful in understanding the political dynamics over the Internet's content.

WHY IS THE CONTENT IMPORTANT?

Examining the Internet content is important because it is the actual material that people see, read, listen to, download, watch, and interact with while online. In sum, it is the end-result of all of the activity at the previous three layers; the finished product that most directly affects the majority of Internet users.

Most of the debates taking place in the media and in governmental policy-making circles regarding "Internet issues" are focused on the content that is publicly available in cyberspace. Such issues that are highly prominent on the legislative agenda in the United States, for example, include the regulation of pornographic material, the protection of privacy and personal information, ensuring free speech and First Amendment principles, cracking down on the piracy of copyrighted works, eliminating spam, and much more. These are all issues that have little, if anything, to do with the physical infrastructure of the Internet, its technical standards and protocols, or (to a lesser extent) the software applications that collectively maintain the Internet's operational functionality. What these issues all share in common is that they are far more narrowly concerned with the substantive nature of the material that users ultimately see.

As a consequence, the majority of public attention and governmental efforts focus on cyberspatial content as their target. The Content layer is their main battleground. The average Internet user is not overly concerned with how things technologically function behind the scenes nor do politicians typically become motivated by software engineering principles and seek formal resolutions involving the production of PHP code. It is the end-result which occupies them. Ultimately, the types of content that are available, and if and how they ought to be regulated, are the focus of policymaking in this most highly visible arena.

WHO GOVERNS THE CONTENT?

The content of the material that is available in cyberspace is governed by national governments, private website operators, and ISPs.

National governments have taken an increasingly active in role in attempting to regulate the content of cyberspace. Goldsmith and Wu have argued that since the mid-2000s geography and governmental coercion have regained their fundamental importance as a more bordered Internet has emerged; one that "differs among nations and regions that are increasingly separated by walls of bandwidth, language, and filters." They further argue that such a bordered Internet has many underappreciated values as it reflects top-down pressures from governments and also reflects "bottom-up pressures from individuals in different places who demand an Internet that corresponds to local preferences."[1]

Perhaps the most highly prominent example of how national governments have been decidedly proactive in attempting to regulate Internet content has been over the issue of protecting minors from indecent and sexually explicit material available online. Since the very inception of the first web browser, the United States Congress has been proactive on this front. Its first attempt came in the form of the Communications Decency Act of 1996 (a common name for Title V of the Telecommunications Act of 1996), which sought to regulate the content published on websites that allowed unfettered access.[2] The two provisions that were most contentious included one that would prohibit the "knowing transmission" of indecent material to any recipient under the age of eighteen and the other would prohibit the use of any "interactive computer service" to send or display offensive material in a manner available to a minor—effectively imposing limits to what material could be published on unrestricted websites. The Communication Decency Act (CDA), however, was partially struck down by the Supreme Court in *Reno v. ACLU (1997)* on First Amendment grounds, ruling that the statute "unduly restricted a large amount of speech that adults have a constitutional right to receive and to address to one another."[3]

Congress, still seeking to pass, what Ripley and Franklin have called, some type of Protective Regulatory legislation[4] soon drafted a second effort at regulating indecent material. The Child Online Protection Act (COPA) was an attempt to respond directly to the Court's decision in *Reno*, making only minor modifications to the CDA which would sufficiently address its concerns. COPA provided for criminal and civil penalties for anyone who "in interstate or foreign commerce by means of the World Wide Web makes any communication for commercial purposes that is available to any minor and that includes any material that is harmful to minors."[5] However, COPA also faced legal hurdles based on First Amendment grounds. A federal District Court, and later affirmed by the Appellate Court, concluded that COPA was a content-based regulation of speech, and therefore subject to strict scrutiny. A Court injunction blocked its enforcement almost immediately after its passage in 1998, and the law was later overturned in the case of *Ashcroft v. ACLU (2004)*.[6] Both these pieces of legislation, the CDA and COPA, were broad attempts to prohibit certain forms of material content to be published on websites.

The third attempt by Congress to create a federal law that would regulate sexually explicit material online was designed with an entirely different strategy. Rather than attempting yet again to directly regulate the content of websites, where prior efforts had not sufficiently withstood judicial scrutiny, the Children's Internet Protection Act (CIPA) was passed in 2003 requiring public institutions which received federal funding to install Internet filters that would disallow access to websites that contained indecent material.[7] In other words, CIPA focused on the demand-side (the access of the end-user) as opposed to attempting to regulate once again the supply-side (what websites could and could not publish). The Supreme Court has since upheld CIPA against a constitutional challenge in *United States v. American Library Association (2003).*[8]

For skeptics of governmental authority in cyberspace, such examples are pointed to as proof of the limits of such authority. They argue that because these policies have a difficult time even passing constitutional muster, let alone proving any actual ability to effectively regulate online content, governments have little to no authority. However, this point is contradicted as proof of governments' clear authority to shape the Internet environment in terms of content can be notably demonstrated with the example of Section 230.

Section 230 of the Communications Decency Act provides immunity from liability for providers and users of an "interactive computer service" who publish information provided by others.[9] What this has effectively come to mean is that private website operators are not to be held legally liable for the content posted by their users.

While Section 230 may have seemed relatively innocuous at the time of its creation, its ramifications have been enormous. Social-networking websites like MySpace and Facebook, that allow users to post their own content, could come into existence since the website companies did not have to fear a litany of litigation. Blogging services like Blogger and WordPress could be created allowing people to publish freely to mass audiences, video-sharing sites like YouTube and photo-sharing sites like Flickr and collaborative projects like Wikipedia could all become possibilities, and much more. This entire phenomenon of websites based on user-generated content has become known as "Web 2.0" as it characterizes, in the eyes of many, the contemporary cyberspatial experience for the majority of users—so much so that Time magazine declared its Person of the Year award in 2006, "You"; a direct reference to the significant impact that user-generated Web 2.0 content was having.[10]

Web 2.0 and all of the websites that are based on user-generated content, including several of the most popular and commercially successful websites in cyberspace, would not have been possible without Section 230. The fear of litigation with no liability protection would have undoubtedly chilled many of the aforementioned websites from coming into existence, or at the very least would have produced sites whose content looked far different than the content that is available on the Web today. For this reason, Section 230 is direct evidence of how national governments continue to retain a governing authority over cyberspatial content, insofar as they clearly have the ability to create policies that constrain or enable different types of behavior with intentional effects—in this case, greatly enabling the widespread proliferation of Web 2.0 sites based on user-generated content.

That said, the limits of governmental authority remain very real. One of the major challenges that national governments face in their attempts to regulate cyberspatial content is the jurisdictional dilemma raised by an Internet that is global in scope—and whose competing laws are not often in harmonization with one another. Data havens exist that promise potential clients—porn purveyors, tax evaders, online gambling services, etc.—that data on their servers will be "physically secure against any legal action." A famous example is the commercial enterprise known as HavenCo, which hosts its servers on the "Principality of Sealand," which is literally a tiny abandoned concrete platform located six miles off the coast of the United Kingdom in the North Sea, linking to the Internet via microwave and satellite connections. HavenCo is generally pointed to as evidence of the futility of territorial government and is based on "the commonplace assumption that governments cannot control what happens beyond their borders, and thus cannot control Internet communications from abroad."[11] While this may be an extreme example, it nevertheless suggests how issues of jurisdiction severely complicate regulatory attempts by national governments, particularly when they act unilaterally.

As a result, many governmental attempts at regulation, even when formalized as public policy, are often rendered ineffective and little more than symbolic actions. For example, the CAN-SPAM Act was signed into law in December 2003, establishing the first national standards for the sending of commercial e-mail in an effort to make spam (non-solicited pornographic and marketing messages) illegal.[12] While initially deemed a popular success, it has since encountered tremendous obstacles to achieving the policy's goals of limiting the amount of spam messages sent over the Internet.

The CAN-SPAM Act was initially introduced as Senate bill S.877 by Conrad Burns (R-MT) along with 22 co-sponsors in April 2003. Coming on the heels of the National Do-Not-Call-Registry, which sought to limit non-solicited telephone marketing calls and had tremendous popular support, the bill passed through the Committee on Commerce, Science, and Technology by July of that same year with little opposition. Within only a few months, the Senate voted in favor of the bill by a vote of 97–0, and the House subsequently voted in favor by a vote of 392–5. By December 2003, President Bush signed the CAN-SPAM Act into law amid an enormous degree of consensus among legislators.[13]

The Act permitted e-mail marketers to send unsolicited commercial e-mail as long as the message contained all of the following four elements: an opt-out mechanism, a functioning return e-mail address, a valid subject line indicating it is an advertisement, and the legitimate physical address of the sender. The Act requires the Federal Trade Commission (FTC) to enforce its provisions; however, other actors involved in the implementation process also include the Senate Committee on Commerce, Science, and Technology, which is charged with oversight, as well as the FCC, which must promulgate the specific rules affecting mobile telephone service messages. Furthermore, the Act sets out both civil and criminal penalties for failing to include any of the above four mandatory elements, as well as other common spamming practices such as harvesting, dictionary attacks, Internet Protocol spoofing, or using open mail relays. It does not allow individuals to sue spammers, but only the FTC, State Attorneys General or corporate ISPs may do so. The Act also preempts any existing state antispam laws.[14]

The role of the FTC is paramount in terms of policy implementation. The Act stipulates that the FTC must submit to Congress an annual report on the effectiveness and enforcement of the Act, devise a system of rewarding informants, create a system requiring spam to be identifiable from its subject line, and provide a specified timetable for establishing a Do-Not-Email Registry (based on the model set by the National Do-Not-Call Registry). Clearly, these provisions are indicative of a strict top-down approach, as described by Pressman and Wildavsky,[15] focusing exclusively on the substantive content of messages sent and received and occurring completely at the federal level.

Since the CAN-SPAM Act went into effect in January 2004, spam on the Internet has risen exponentially. In 2003, the average daily volume of spam was 15 billion messages, as compared to 164 billion in 2008. In 2003, 45 percent of all e-mail sent was spam, as compared to 96.5 percent in 2008. In 2003, the amount of money spent battling spam annually was $20.5 billion, as compared to $140 billion in 2008. Furthermore, spam has taken on a more destructive quality in the years since the Act's passage when the issue was focused on unsolicited commercial marketing messages, thanks to e-mail attachments that link to websites that infect computers with malicious code. Spammers have also become more brazen in attempting to steal data or take control over the infected computer and join it to botnets for future attacks.[16]

Implementation challenges associated with the CAN-SPAM Act have been evident from the outset of the policy's adoption. First, it is an immense problem to identify and locate spammers, given an assortment of programming techniques such as automated dictionary attacks and IP spoofing. Second, even once the spammer is located it can often be extremely difficult to develop sufficient evidence to prove the spammer is legally responsible for actually sending the spam, again given an assortment of programming techniques through the use of proxy servers and subnet masking. Lastly, there is the jurisdictional dilemma posed by the Internet's global dimension. For instance, if a small firm sends spam from a computer in the United States, transmitted through a mail server located on an offshore island nation without any antispam laws, and received by a citizen of a foreign nation outside of U.S. sovereignty, determining which authority has proper jurisdiction is often highly disputed.[17]

The example of the CAN-SPAM Act is indicative of the limits of governments when their legislative policy solutions fail to address the technology itself. E-mail operates through a technical protocol called Simple Mail Transfer Protocol (SMTP), created by the IETF. SMTP specifies that a minimum of four logical computing components must be used in the e-mail process—the sender's, the recipient's, and most crucially, a mail server located at the ISP of both the sender and recipient. When a mail server transmits a message to another mail server over the Internet, the message must have a "header" which contains lines of information that provide details about the message, the sender, and the transmission. It is through manipulation of these headers that spammers are able to conceal their location and identity by falsifying the information contained in the headers. The SMTP protocol facilitates spam because it does not require accurate routing information except for the intended recipient of the e-mail, and performs no authentication.[18]

Thus, the FTC's final report to Congress in 2004[19] concluded that due to the technical nature of SMTP and other concerns regarding proposed policy alternatives,[20] the creation of a National Do-Not-Email Registry "in any form would not have any beneficial impact on the spam problem." The FTC further stated that a viable authentication standard is not only required to make such a registry effective, but "may even substantially address the underlying [spam] problem" itself.

However, the problem of spam has, in fact, been partially mitigated in recent years from the end-user's perspective; not because of direct government intervention or as the result of formal policymaking, but through private-sector initiatives. It has been private website operators, acting in their own private commercial interests, who have greatly reduced the amount of spam that ultimately reaches their users' inboxes. Today's antispam tools catch anywhere from 95 percent to 98 percent of spam before it enters ISP or corporate networks, and 71 percent of Internet users are now protected by spam filters.[21] They have accomplished this feat through technological innovations such as improved filtering algorithms, Zero-Hour systems, and features that allow users to flag spam messages as a form of notification.[22]

What CAN-SPAM demonstrates is that policies aimed at addressing technological problems ought to seek technological solutions to work in conjunction with purely legislative ones. Is it possible to better regulate spam? Many experts within the scientific and engineering community generally agree the answer is yes.[23] Public policies which aim at regulating the content of cyberspace would be most effective if they were designed to address the other various Internet layers that have been described. In this example, spam could be best mitigated by seeking to influence the structural design of the SMTP protocol, while simultaneously encouraging the private sector to innovate better filters and other software solutions, and by governments establishing strict criminal and civil penalties for offenders. As the FTC concluded, an authentication mechanism, most easily built into the header specifications of the protocol, would facilitate an effective means of locating the origin of spam messages.[24] Similar types of efforts have already proven successful at accomplishing this, such as the deployment of digital certificates providing developers with a means of authenticating commercial sales transactions over the Web, thereby contributing to the rapid expansion of e-commerce services in the late 1990s.

Thus, national governments certainly play an active and significant role in regulating the content of cyberspace, as demonstrated by Section 230, as do private website operators who often, in their own private interest, implement their own policies in the form of technological solutions to technological problems. As was the case with spam, this approach often succeeds where governmental policymaking, on its own, fails.

However, there is another major actor in the governance equation when it comes to content. Increasingly, ISPs are demonstrating their own authority to constrain and enable different types of cyberspatial behavior with intentional effects.

How ISPs have authority over cyberspatial content is illustrated by the case of P2P file-sharing. In 1999, a college student named Shawn Fanning wrote a software application, called Napster, which posed an immediate challenge to both the business model of the music industry as well as national copyright law. Using Napster, people could easily search for and download music files from people's hard drives around the world, with Napster simply providing a centralized list of what content was available. Soon, 26.4 million verified users were on Napster trading billions of songs at its peak in February 2001.[25]

It wasn't long before federal courts in California concluded that Napster was a "contributory infringer" of copyright in the case of *A&M Records, Inc. v. Napster, Inc. (2001)*,[26] and consequently ordered it to be shut down. This shutdown was possible on a technical level for two reasons. First, its directory of music files was centrally located, meaning that shutting down the central server would effectively lead to a total system collapse. Second, it was geographically located exclusively within the United States.[27]

When Napster shut down in 2001, file-sharing users sought alternatives. Gnutella networks were the most obvious successor, and were able to maintain their functionality because, unlike Napster, they had a radically decentralized "peer-to-peer" architecture with no central directory of files. Other successful file-sharing alternatives soon arose, like LimeWire and Kazaa, based on a similar type of decentralization and were headquartered strategically overseas. By early 2004, Kazaa became the most downloaded piece of software in history, having been downloaded 319 million times. By 2005, over 57 million Americans were using file-sharing software—more than the number who voted for President George W. Bush.[28]

While new legal challenges were brought by the RIAA and the MPAA against these P2P software vendors in the courts, the industry began pursuing a new tactic of suing individual users of the software. By January 2009, lawsuits had been brought against nearly 35,000 Americans;[29] however their relative level of effectiveness is in doubt. The number of lawsuits paled in comparison to the number of actual file-sharers, the number of file-sharers soon rose exponentially and the software became more popular than before the lawsuits began,[30] and the courts soon took issue with the legality of the lawsuits themselves citing the industry's very fallible method of identifying possible copyright infringers based on IP addresses, which is not necessarily a reliable means of identification in cyberspace. The strategy was repeatedly justified by the RIAA, however, claiming "the real point of the suits was not to eliminate filesharing but to marginalize it and thus prevent companies like Kazaa from becoming mainstream, legitimate businesses, and real competitors to the labels."[31]

The next central legal challenge came in the case of *MGM Studios, Inc. v. Grokster, Ltd. (2005)*.[32] Although Kazaa representatives and others did not show up for the lawsuit in order to avoid the enforcement powers of U.S. authorities, remaining defendants based their argument on the Supreme Court's ruling in the famous *Sony Betamax* case of 1984.[33] In it, the Court ruled that it is indeed legal to create standardized technologies (like a VCR or photocopier) that may be potentially used for copyright infringement so long as they also were capable of "substantial noninfringing uses." In April 2003, a federal District Court dismissed the *Grokster* case citing the *Sony Betamax* decision, and the Ninth Circuit Court of Appeals later upheld that decision acknowledging the legitimate and legal uses of P2P software. However, when the Supreme Court ruled on the case in 2005, it unanimously reversed the lower courts in *Grokster* and declared the business model of the file-sharing firms illegal because they had, with obvious intent, "induced" users to break the law.

The Court's decision sparked the next phase in the file-sharing narrative. Once again, new adapted technologies quickly arose to succeed Kazaa and its commercial brethren, the most prominent among them called BitTorrent. BitTorrent was another P2P communications protocol used for distributing large amounts of data and by some estimates accounted for as much as 55 percent of all Internet traffic, as of February 2009.[34] Thus, despite the efforts of governmental institutions and the music industry, file-sharing remains a significant challenge to established real-space authorities.

This brings us back to the increasing role of ISPs in the governance debate. In 2007, investigations revealed that Comcast was looking at its users' web traffic and secretly blocking BitTorrent uploads to users outside Comcast's network. The EFF alleged that Comcast blocked BitTorrent with a classic hacker technique known as "spoofing," where the hacker poses as someone he isn't, in this case another user. It is "as if he and I were having a phone conversation, and then halfway through Comcast interrupts us and in my voice tells him to hang up, and in his voice tells me the same thing."[35] Comcast and other ISPs were revealed to have started sniffing out P2P traffic on their networks and curbing it, either slowing file-sharing to a trickle or bringing it to a halt.[36]

The FCC ultimately intervened and ruled with a bipartisan majority to require Comcast to stop this ongoing practice and also disclose all of its network management practices.[37] The case proved to be a focusing event sparking a debate over the issue of Net Neutrality—the principle that all data on the Internet ought to be treated equal regardless of content, at least in terms of transmission speeds and pricing.[38]

However, ISPs are nevertheless increasingly being turned to as agents of policymaking and implementation in matters of regulating Internet content. Some ISPs, such as Sprint, Time Warner, and Verizon, agreed to block websites that contain child pornography after forming a deal with New York Attorney General Andrew Cuomo.[39] Several bills have been introduced in the House and Senate that call for ISPs to keep logs about their users' Internet activities for two years to aid police investigations.[40] Also, as it relates specifically to file-sharing, the RIAA announced its intention in late 2008 to shift its strategic focus away from bringing lawsuits against individual file-sharers and instead use ISPs as their official conduits. Placing a notable emphasis on college campuses, their efforts have since been geared toward altering the policy environment so that when illegal file-sharing or downloading is detected, the ISP, not the RIAA, "will contact the culprit via e-mail, requesting they stop after the first of two warnings. If the user does not stop after the second warning, the ISP will slow down service or cut it off all together."[41] ISPs are warranting this renewed attention because they have demonstrated their ability, through policymaking, to both constrain and enable different types of cyberspatial behavior due to their unique position as being the gatekeeper for people's access to the Internet.

Thus, national governments, private website operators, and ISPs all play a significant role in governing the content of the Internet, despite their respective limitations.

HOW ARE POLICIES BEING MADE AT THE CONTENT LAYER?

TOS Agreements are the rules that users must agree to abide by in order to use a service, and are the de facto law of, what scholars like Jonathan Zittrain refer to as, the Web's various "walled gardens."[42] These TOS Agreements have become an important form of policy themselves that often have dramatic effects on constraining or enabling different types of cyberspatial behavior.

Considering the triumvirate of governing actors established in the previous section—national governments, private website operators, and ISPs—it is TOS agreements that clearly are the policies of record for the latter two. Private website operators, particularly of those that are commercial in nature, rely on TOS agreements to establish behavioral norms for its users' activities such as what content is acceptable to publish, post copyright notices, and lay out the company's marketing policies. TOS agreements also stipulate the penalties for violating the website's rules, which includes the criteria for total expulsion.

Because each website is administered by its own private operator, and is therefore a private space despite its often public accessibility, the private operators can create the TOS for their sites however they choose so long as it does not violate the territorial laws in which the company is incorporated, or in which their servers are geographically located. Private website operators are increasingly using TOS agreements as a form of policymaking which sometimes even supersedes the policymaking capabilities of governments, particularly in the transnational context.

For example, in February 2009, Facebook made a subtle change to its TOS agreement—the type that ordinarily goes unnoticed. The new TOS included the following phrase:

> You hereby grant Facebook an irrevocable, perpetual, non-exclusive, transferable, fully paid, worldwide license (with the right to sublicense) to (a) use, copy, publish, stream, store, retain, publicly perform or display, transmit, scan, reformat, modify, edit, frame, translate, excerpt, adapt, create derivative works and distribute (through multiple tiers), any User Content you (i) Post on or in connection with the Facebook Service or the promotion thereof subject only to your privacy settings or (ii) enable a user to Post, including by offering a Share Link on your website and (b) to use your name, likeness and image for any purpose, including commercial or advertising, each of (a) and (b) on or in connection with the Facebook Service or the promotion thereof.[43]

This particular change to the TOS did not go unnoticed. A popular blog, The Consumerist, informed the world about it, running a headline, "Facebook's New TOS: 'We Can Do Anything We Want With Your Content. Forever'."[44]

> Facebook's terms of service (TOS) used to say that when you closed an account on their network, any rights they claimed to the original content you uploaded would expire.
>
> Not anymore.
>
> Now, anything you upload to Facebook can be used by Facebook in any way they deem fit, forever, no matter what you do later. Want to close your account? Good for you, but Facebook still has the right to do whatever it wants with your old content. They can even sublicense it if they want.

The story snowballed and spread in a classically viral manner. The Consumerist post received hundreds of comments, thousands of Diggs, and half a million page views. Meanwhile, outraged Facebook users created several groups on the site itself with names like "People Against the New TOS," each one quickly recruiting tens of thousands of members.[45] Finally, the mainstream media caught on, with the story reported on Fox News and in the New York Times, among others, ultimately pressuring the company to revert to its previous policy.[46]

While this is a case of Facebook giving in to pressure and ultimately altering the terms of their TOS Agreement, what this story truly illustrates is that, although its users and the mainstream media mobilized to influence Facebook's policy, it was still entirely up to Facebook whether or not to do so. The company had a clear authority to create the policies that would govern its own private website, and this power was embodied through their TOS agreement. Again, the distinction between who has influence versus who has authority is paramount in determining governance.

Of course, the use of TOS agreements to regulate customer or member behavior is not unique to cyberspace. It is a long-established practice that has been followed by various member-based communities and private commercial firms that provide assorted services—for example, credit card companies. What is notable, however, is that TOS agreements on the Web may have a larger impact on affecting behavior than their real-space counterparts insofar as they actually create the virtual environments in which interactions are occurring. Particularly when the regulatory limits of traditional governmental institutions are evident in the transnational context, TOS agreements that govern the specific private spaces of the Web are, in effect, the de facto law.

This point is further supported by examining how it is not only private website operators that use TOS agreements as a form of policymaking, but ISPs as well. As the gatekeepers to Internet access, ISPs have repeatedly used TOS agreements to regulate what activities are, and are not, possible for their users. For example, as already cited, commercial ISPs have altered their TOS agreements to enable them to manage their network traffic in ways which disallow file-sharing and put bandwidth caps in place that make private website hosting problematic for residential users. Noncommercial ISPs, such as public universities, have also used TOS agreements to cover issues including protection from security threats, setting up wireless hotspots for sharing connections, and emulating commercial environments for testing software applications.[47]

When it comes to how governments are creating policies at the Content layer, the answer can best be explained using Heclo's theory of Issue Networks.[48] Based on the premise that closed Iron Triangles are "disastrously incomplete" because they miss the fairly open networks of people that also interact in important ways with government officials, issue networks include many disparate actors whose webs of influence guide the exercise of power. Participants move in and out of the networks constantly and operate on many levels. Powerful interest groups and knowledgeable individuals alike are often represented, and the true experts in the networks are those who are "issue-skilled" regardless of formal professional training. As Heclo described, an issue network is a "shared-knowledge group" having to do with some aspect of public policy.

> [Issue networks are] therefore more well-defined than, first, a shared-attention group or "public"; those in the networks are likely to have a common base of information and understanding of how one knows about policy and identifies its problems. But knowledge does not necessarily produce agreement. Issue networks may or may not, therefore, be mobilized into, second, a shared-action group (creating a coalition) or, third, a shared-belief group (becoming a conventional interest organization). Increasingly, it is through networks of people who regard each other as knowledgeable, or at least as needing to be answered, that public policy issues tend to be refined, evidence debated, and alternative options worked out—though rarely in any controlled, well-organized way.[49]

Because of the diversity of policy debates occurring at the Internet's Content layer, issue networks help explain the fluid range of participants. Each individual issue, whether the regulation of pornographic material, spam, file-sharing, or others, has an identifiable subgroup of actors significantly affecting the policy debate—and those groups, and their dynamics, are particular to each issue. For instance, the issue network involved in the regulation of spam is heavily dominated by interests such as the U.S. Chamber of Commerce, who seek to maintain the legitimacy of certain types of commercial solicitations by e-mail, software companies like Microsoft and Yahoo, whose mail servers must handle the flood of spam messages, Congress, who creates the affecting legislation, and the FTC, who is charged with enforcement. Contrast this with the issue of file-sharing which is dominated by a largely different set of actors and comprising of a different power dynamic—the file-sharing software firms are not notably influential on the domestic front since they operate primarily overseas, however advocacy groups such as the EFF influence the debate through mobilization efforts and litigation, ISPs attempt to do so through TOS agreements and bandwidth caps at a more technical level, the RIAA represents the music industry, and the U.S. federal court system has been the primary arena for resolving disputes and guiding governmental policy, more so than Congress.

What this demonstrates is that, because the Web consists of so much content, and content of such diverse types, to attempt to define a single model for policymaking would be to oversimplify the power dynamics that actually occur at the Content layer. Issue networks explain the manner in which governments are creating policies in that they account for the enormous range of activities occurring in public policy as it relates to cyberspatial content. How policies are being made regarding the regulation of pornographic material is inevitably going to be different than the manner in which policies are being made regarding bandwidth caps and Net Neutrality. Different issues involve different debates and different interests, and understanding that the policy networks will not always be homogenous is critical in understanding how policies are being made by governments at the Content layer, in general.

NOTES

1. Goldsmith and Wu VII-VIII.
2. *Public Law 104–104* (February 1996). Retrieved on February 21, 2013 from http://thomas.loc.gov/cgi-bin/query/z?c104:S.652.ENR:.
3. Reno v. ACLU, no. 521, Supreme Ct. Of the US, 1997.
4. Ripley and Franklin's typology of public policies—distributive, redistributive, protective regulatory, and competitive regulatory policy types. See R. B. Ripley and G. A. Franklin, *Bureaucracy and Policy Implementation* (Homewood, IL: Dorsey Press, 1982).
5. *Children's Online Protection Act* 47 USC s231 (1998). Retrieved on February 21, 2013 from http://www.law.cornell.edu/uscode/text/47/231.
6. Ashcroft v. ACLU, no. 124, Supreme Ct. of the US, 2004.
7. *Children's Internet Protection Act* (2000). Retrieved on February 21, 2013 from http://thomas.loc.gov/cgi-bin/query/z?c106:S.97.IS:.
8. United States v. American Library Association, no. 02–361, Supreme Ct. of the US, 2003.
9. *Telecommunications Act of 1996* 47 USC Title 47, Chapter 5, Subchapter II, Part I, s230. Retrieved on August 3, 2009 from http://www.law.cornell.edu/uscode/47/230.html.

10. Lev Grossman, "Time's Person of the Year: You," *Time* December 13, 2006. Retrieved on August 3, 2009 from http://www.time.com/time/magazine/article/0,9171,1569514,00.html.

11. Goldsmith and Wu, 65–66.

12. *Public Law 108–187* (December 2003). Retrieved on February 21, 2013 from http://uscode.house.gov/download/pls/15C103.txt.

13. *Bill Summary and Status, 108th Congress.* Retrieved on February 21, 2013 from http://thomas.loc.gov/cgi-bin/bdquery/z?d108:S.877:.

14. *Bill Summary and Status, 108th Congress.*

15. Pressman and Wildavsky used a top-down approach to analyze "implementation deficits" which are defined by the gap between policy promises and performance, and highlight the difficulties in translating broad policy agreements into specific decisions over "technical details." See Jeffrey Pressman and Aaron Wildavsky, *Implementation,* 3rd ed. (Berkeley, CA: University of California Press, 1984).

16. Carolyn Duffy Marsan, "CAN-SPAM: What Went Wrong?," *Network World* October 6, 2008. Retrieved on August 3, 2009 from http://www.networkworld.com/news/2008/100608-can-spam.html.

17. Derek E. Bambauer, "Solving the Inbox Paradox: An Information-Based Policy Approach to Unsolicited E-mail Advertising," *Virginia Law Review* 10.5 (Spring 2005). Retrieved on February 21, 2013 from http://www.vjolt.net/vol10/issue2/v10i2_a5-Bambauer.pdf.

18. Federal Trade Commission, *National Do Not Email Registry: A Report to Congress* (June 2004).

19. Federal Trade Commission, *National Do Not Email Registry: A Report to Congress.*

20. Concerns over security and privacy, from a constitutional perspective, were discussed as they related to proposed alternative models such as a national registry of individual e-mail addresses, a registry of domains, and a registry of e-mail addresses using third-party forwarding service.

21. Marsan.

22. Elinor Mills, "Postini: Google's Take on E-mail Security," *CNET* July 1, 2009. Retrieved on August 3, 2009 from http://news.cnet.com/8301–1009_3–10276548–83.html.

23. The IEEE has, in the past, proposed new spam regulations. See Park So Young, "Proposal of a New Effective Spam Mail Regulation," *Advanced Communication Technology, ICACT 2005*, The 7th International Conference (2005). Retrieved on February 21, 2013 from http://ieeexplore.ieee.org/xpl/articleDetails.jsp?reload=true&arnumber=1462971&contentType=Conference+Publications.

24. Dennis Callaghan, "FTC Shoots Down Spam Registry, Boosts Authentication Scheme," *EWeek* June 16, 2004. Retrieved on February 21, 2013 from http://www.eweek.com/c/a/Messaging-and-Collaboration/FTC-Shoots-Down-Spam-Registry-Boosts-Authentication-Scheme/.

25. Jupiter Media Metrix, "Global Napster Usage Plummets, But New File-Sharing Alternative Gaining Ground," July 20, 2001. Retrieved on August 3, 2009 from http://www.comscore.com/press/release.asp?id=249.

26. A&M Records, Inc. v. Napster, Inc., 239 F.3d 1004 (9th Cir. 2001). Retrieved on August 4, 2009 from http://cyber.law.harvard.edu/~wseltzer/napster.html.

27. Goldsmith and Wu, 108.

28. Goldsmith and Wu, 108–115.

29. Jeanine Budd, "RIAA Backs Away from Suing Students, Begins Targeting ISPs," *Gateway* January 20, 2009. Retrieved on August 5, 2009 from http://media.www.unogateway.com/media/storage/paper968/news/2009/01/20/News/Riaa-Backs.Away.From.Suing.Students.Begins.Targeting.Isps-3590227.shtml.

30. Janko Roettgers, "Limewire Wants to Give Record Labels a Cut of Its Ad Revenue," *Electronic Frontier Foundation P2P Blog* May 13, 2008 (noting LimeWire has 80 million users generating about 5 billion search requests every month); Ernesto, "BitTorrent Trio Hit a Billion Pageviews a Month," *TorrentFreak* June 11, 2008 (describing three BitTorrent websites—Mininova, The Pirate Bay, and isoHunt—that have entered the list of top 100 most visited websites on the Internet); "Azureus Announces One Million Unique Visitors to Its Digital Media Platform Currently Code Named Zudeo," *BNET Business Wire* February 16, 2007 (boasting that Azureus is "the provider of the most popular P2P application for the transfer of large files" and citing over 140 million downloads of its application in the past few years).

31. Goldsmith and Wu, 115.

32. MGM Studios, Inc. v. Grokster, Ltd. 545 U.S. 913 (2005). Retrieved on August 4, 2009 from http://caselaw.lp.findlaw.com/cgi-bin/getcase.pl?court=US&navby=case&vol=000&invol=04–480.

33. Sony Corp. of America vs. Universal City Studios, Inc., 464 U.S. 417 (1984).

34. "BitTorrent Still King of P2P Traffic," *TorrentFreak*. Retrieved on August 4, 2009 from http://torrentfreak.com/bittorrent-still-king-of-p2p-traffic-090218/.

35. David Downs, "BitTorrent, Comcast, EFF Antipathetic to FCC Regulation of P2P Traffic," *San Francisco Weekly* January 22, 2008. Retrieved on August 3, 2009 from http://www.sfweekly.com/2008–01–23/news/bittorrent-comcast-eff-antipathetic-to-fcc-regulation-of-p2p-traffic/.

36. Michael Calore, "As Traffic Grows, P2P Vendors Seek Peace with ISPs," *Wired.com* August 30, 2007. Retrieved on August 5, 2009 from http://www.wired.com/software/webservices/news/2007/08/p2p.

37. Federal Communications Commission, *Memorandum Opinion and Order, FCC-08–183A1* (August 20, 2008). Retrieved on August 4, 2009 from http://hraunfoss.fcc.gov/edocs_public/attachmatch/FCC-08–183A1.pdf.

38. The Net Neutrality debate will be explored in greater detail in the final chapter.

39. Danny Hakim, "Net Providers to Block Sites with Child Sex," *New York Times* June 10, 2008. Retrieved on August 4, 2009 from http://www.nytimes.com/2008/06/10/nyregion/10Internet.html.

40. Declan McCullagh, "Bill Proposes ISPs, Wi-Fi Keep Logs for Police," *CNET* February 19, 2009. Retrieved on August 5, 2009 from http://news.cnet.com/8301–13578_3–10168114–38.html.

41. Jeanine Budd, "RIAA Backs Away from Suing Students, Begins Targeting ISPs," *Gateway* January 20, 2009. Retrieved on August 5, 2009 from http://media.www.unogateway.com/media/storage/paper968/news/2009/01/20/News/Riaa-Backs.Away.From.Suing.Students.Begins.Targeting.Isps-3590227.shtml.

42. Jonathan Zittrain, *The Future of the Internet—And How to Stop It* (New Haven, CT: Yale University Press, 2008).

43. *Facebook Terms of Service*. Retrieved on February 18, 2009 from http://www.facebook.com/legal/terms.

44. Chris Walters, "Facebook's New Terms of Service: We Can Do Anything We Want With Your Content. Forever," *The Consumerist* February 15, 2009. Retrieved on February 18, 2009 from http://consumerist.com/5150175/facebooks-new-terms-of-service-we-can-do-anything-we-want-with-your-content-forever.

45. "People Against the New Terms of Service (TOS)," *Facebook Group*. Retrieved on February 18, 2009 from http://www.facebook.com/group.php?sid=a6cdf0abf38c1d67123c77fc196e546c&gid=77069107432.

46. Brad Stone and Brian Stelter, "Facebook Withdraws Changes in Data Use," *New York Times* February 18, 2009. Retrieved on February 18, 2009 from http://www.nytimes.com/2009/02/19/technology/Internet/19facebook.html?_r=3&hp.

47. Penn State University, *ISP Service for University Campus Locations*. Retrieved on August 6, 2009 from http://its.psu.edu/policies/ispunivtesting.html.

48. Hugh Heclo, "Issue Networks and the Executive Establishment," *Public Policy: The Essential Readings*, ed. Stella Z. Theodoulou and Matthew A. Cahn (Upper Saddle River, NJ: Prentice-Hall, 1995), 48.

49. Heclo, 48.

Part III

CASE STUDY: U.S. NATIONAL CYBERSECURITY POLICY, POST-9/11

Chapter 7

Analysis of the National Strategy to Secure Cyberspace

In Part I, we developed a new four-layer conceptual model and determined governance of the Internet based on the political architecture that it produced. Our argument is that this layers-based approach can be a helpful tool for conceptualizing Internet issues—both their problem and solution streams.

Now, let's put that argument to the test. In order to demonstrate how our four-layer model can be helpful in understanding complicated and often misunderstood Internet issues, let us apply the model to a detailed case study on U.S. national cybersecurity policy. Specifically, we will apply the four-layer model in order to test the following hypothesis—the commonly cited view that cybersecurity policy's failures are the result of a flawed policy design that focuses almost exclusively on voluntary public-private partnerships.

Cybersecurity serves as a meaningful test case for several reasons. First, it is an issue of vital importance with rapidly growing consequences to national and global economies. As the U.S. Director of National Intelligence told the Senate Intelligence Committee in February 2013, the threat of cyberattacks was more pressing than the risk of an attack by global terrorist networks.[1] It is an issue that highlights both the challenges that policymakers face in a global jurisdictionally challenged cyber environment as well as one that serves as a model example for how the unique dynamics of all four Internet layers may, or may not, be formally addressed through public policy in the pursuit of political objectives.

Second, it is highly complex both in technical and political terms. Its technical dimension focuses on enhancing security primarily through code, the details of which are often incomprehensible to government legislators, and this is coupled with a political calculus that must account for numerous governing actors, each with a level of authority that is relative to the particular Internet layer in question, and where competing interests are structurally decentralized. This complexity involving all the associated variables that must be taken into account means that lessons about how technical solutions can achieve political outcomes on the Internet might be applicable to other policy issue areas as well.

Third, it's relatively new. Like the Internet itself, national cybersecurity policy, at least in some form, can trace its early origins back several decades, however U.S. national cybersecurity policy as it exists today really took shape following the terrorist attacks of September 11th, 2001. Therefore, it is still in its nascent stages and we can, as scholars, observe its evolutionary development practically from its inception, and that is something we hope will prove instructive.

Fourth, and most importantly from a research perspective, there is an actual concrete policy to examine. There are many possible case studies of Internet issues that can be considered for analysis, however, the goal of this project from the outset has been to draw conclusions, not about specific issue areas, but about governance—and governance, according to our stated definition, is demonstrated through the creation of *policies* that can constrain or enable behavior with intentional effects. Thus, in order to test the applicability of our four-layer conceptual model, we need to utilize a policy-based approach, which makes it crucially necessary to have a concrete policy—and underlying policymaking process—to examine. Cybersecurity serves this purpose with its central guiding policy—the Bush administration's *NSSC*.

Our goal in the following section is to test our four-layer conceptual model by analyzing U.S. national cybersecurity policy in terms of the Internet's four layers. As the following pages will demonstrate, even with an issue as comprehensive as cybersecurity, both the problem stream and the solution stream can be framed in terms of how policy affects, and is affected by, the Internet's Infrastructure, its Technical Protocols, its Software Applications, and its Content. Furthermore, our map of the Internet's political architecture, where we identified the primary holders of governing authority at each layer, also proves valuable in understanding existing U.S. cybersecurity policy's focus on specific actors and industries.

By applying the lens of our four-layer model, we will be able to frame both problem stream and solution (or policy) stream characterizing the U.S. cybersecurity issue in more clear terms:

As for the problem stream, the primary threat is the destruction of physical infrastructure and the hijacking of core industrial systems occurring at the Infrastructure Layer; at the Applications Layer, the threat is the infiltration of web application software on both client- and server- sides; and a the Content Layer, the threat that is to be defended against is the defacement of websites or the possibility of them being shut down completely.

As for the solution stream, the policies that are designed to address those threats include, at the Infrastructure Layer, that private-sector firms are encouraged to submit vulnerability information to the DHS (with the ability to do so confidentially); at the Applications Layer, that private-sector firms are encouraged to voluntarily deploy technical measures for detection and analysis (using assorted software applications and network management tools); and at the Content Layer, that website operators and ISPs are encouraged to put measures in place to detect cyberattacks, mitigate their effects, and notify others of the threat. DHS and its subagencies will coordinate efforts among the private stakeholders and act as the central hub for information-sharing. Thus, as will be illustrated, the four-layer model proves a helpful conceptual tool in understanding this important policy issue area.

The narrative of U.S. cybersecurity policy can really be thought of in two parts. First, in the initial years following the terrorist attacks of September 11th, 2001, the story is about the policymaking process that eventually led to the *NSSC* policy document. Second, in the years since, the story is about the formation of a new bureaucratic regime headed by the U.S. DHS.

We will begin, first, by applying the four-layer model to conceptualizing the problem definition. The generalized problem which U.S. cybersecurity policy is designed to address—namely, digital threats to the nation's critical cyber assets—can be made more specific by deconstructing the problem using a layers-based approach. What we will find is that the major threats to national cybersecurity occur primarily at the Infrastructure and Application layers. Furthermore, we will analyze the categorical and specific mechanisms by which threat agents pursue their goals at each of the aforementioned layers. This will be accomplished by introducing a new typology that draws important distinctions between cyberterrorism, hacktivism, cracktivism, and cyberwarfare, and places specific deployment mechanisms like viruses, worms, botnets, and DDOS attacks in this context. After framing the problem definition in this manner, we will seek to determine its consequences within the framework of our political architecture and ascertain the extent to which the U.S. government has the governing authority to create effective cybersecurity policies.

Second, we will examine the policymaking process that led to the primary document currently guiding U.S. national cybersecurity policy—the Bush administration's *NSSC*. This process will be characterized as open, but flawed. Under the Bush administration, and with Richard Clarke as the central guiding figure, a presidential advisory board was established and released 53 questions to the public for comment. The Board then drafted an initial proposal which was discussed in numerous town hall meetings across the country, ultimately leading to the final draft version of the policy. This process, while open to the public, was heavily influenced at every stage by large private corporations.

Third, we will perform a detailed analysis of the policy design behind the *National Strategy*. The policy design of this document is important in its implicit recognition of all four layers in our conceptual framework. It calls for enhancing the protection of the nation's critical cyber assets by bolstering the defenses of the physical infrastructure, and directly references how this can be achieved through designing more secure technical standards and protocols, promoting more secure software application development in the private commercial sector, and by patrolling Web content. It is here where we will also test the hypothesis that the policy's failures are attributable to a design which relies too heavily on voluntary public-private partnerships.

Fourth, we will study the policy's implementation. From the outset, the federal government came under heavy criticism for failing to allocate enough resources to the problem and for not going beyond the voluntary public-private measures prescribed by the *National Strategy*. Implementation was clearly hindered by, initially, this lack of adequate resources as well as by a high turnover rate at the top levels of the newly created executive bureaucracy—the DHS's NCSD. Subsequently, implementation was made even more problematic by confusion stemming from organizational conflict among numerous federal agencies, competing vigorously for authority.

This characterization remains largely in place, thus we will next seek to clarify the current bureaucratic regime governing U.S. national cybersecurity policy in the Obama administration. As will be explained, this regime, which had been headed by the NCSD division within the DHS, first experienced a weakening of its authority when its primary role came into conflict with that of the newly created NCSC. Under the Obama administration, the *National Strategy* has remained the seminal policy document on the issue and has already exhibited signs of becoming path dependent. The NCSD has found itself competing intensely to retain its governing authority in the face of challenges brought by the DoD and the NSA, and particularly the military's CYBERCOM command center. Meanwhile, in Congress, the Senate Committee on Homeland Security and Governmental Affairs, the House Select Committee on Intelligence, and the House Homeland Security Committee have become leading actors on the issue.

Finally, after applying our four-layer model to analyzing the problem definition, policymaking process, policy design, and policy implementation, we will then attempt to tie all of this together by examining cybersecurity policy in action—namely, what actually happens in the face of a cyberattack. This analysis will provide evidence of three key points that, we argue, can be applied to other Internet issue areas as well: first, the centrality of the private sector, particularly in preventing attacks; second, the reliance on software applications and technical protocols both in prevention and response, particularly network-monitoring tools and other specific software products; and third, that the federal government's role is relegated primarily to being a coordinator among the numerous private actors identified in our political architecture of the Internet, each possessing a level of governing authority in their own right.

In this manner, as we shall see, the four-layer model assists in conceptualizing both the problems associated with cybersecurity threats as well as the policies that have been designed and implemented to face them. Our argument is that U.S. national cybersecurity policy's overriding design and policymaking process are indeed reflective of how all four conceptual layers are important in their own right, and that this confirms the utility of the four-layer model in general. Additionally, in refuting the hypothesis under examination, we will argue that the acknowledged failures of U.S. cybersecurity policy have more to do with an implementation process characterized by organizational turmoil within the Executive Branch of the federal government than with a flawed policy design or policymaking process—and, in fact, this only serves to reinforce our argument that government alone does not have adequate governing authority to achieve their desired outcomes. Finally, in addressing this common critique of the policy—that it relies too heavily on public-private partnerships—we will assert that this is not so much a policy preference or design choice, but a matter of necessity; a recognition of the Internet's decentralized technical reality where numerous governing actors have authority at different layers. Thus, the lessons of U.S. national cybersecurity policy reaffirm that, on the Internet, political objectives are most attainable by targeting policies at the layer most appropriate for specific problems, or by targeting one layer in order to intentionally produce cascading effects at another layer entirely.

PROBLEM DEFINITION: HOW TO DEFEND THE NATION'S CRITICAL CYBER ASSETS FROM ATTACK?

What problem is U.S. cybersecurity policy meant to address? To use John W. Kingdon's vocabulary, public policy is created in response to a specific need, or a "problem stream."[2] In the case of cybersecurity policy, the problem stream exists in the form of digital threats to the nation's critical cyber assets—and in the context of our four-layer model, this means threats to the Internet's physical infrastructure, security vulnerabilities in widely used web software applications, and destructive attacks on the Web's content.

There is a litany of terms referring to cybersecurity threats like hacking, cracking, cyberterrorism, and cyberwarfare that are often thrown around and used interchangeably. From a security perspective, that is a mistake. There are fundamental differences between the various forms of online threats and activities that have emerged, and these distinctions must be made explicit if public policies are to address their associated security challenges.

In practical terms, the greatest cybersecurity threat, or nightmare scenario, facing Homeland Security officials is one where core industrial systems are infiltrated and taken over. The attacker could then wreak havoc over the systems being managed, like a city's water supply, or sensitive data could be stolen. These are Infrastructure-layer threats, deployed by code.

The dangers of this type of threat became clearly apparent when, in 2009, a teenage computer programmer named John Matherly launched a search engine called Shodan that mapped and captured the technical specifications of devices linked to the Internet. Shodan unexpectedly revealed countless numbers of industrial control computers—the systems that automate such things as water plants and power grids—and many were found to be wide open to exploitation by even moderately talented hackers. In various examples, one hacker broke into a water plant south of Houston using a default password he found in a user manual, another Shodan user accessed the cyclotron—a nuclear particle accelerator—at the Lawrence Berkeley National Laboratory, and another discovered thousands of unsecured Cisco routers across the Internet which direct data traffic on the networks.[3]

Disabling core infrastructural components like water supply systems, electrical power grids, or air traffic control systems are deemed as perhaps the single greatest cybersecurity threat. However, the threat of cyberwarfare does indeed work both ways. In 2010, the Stuxnet computer worm managed to infiltrate a number of Iranian nuclear facilities, ultimately destroying nearly 1,000 out of Iran's 6,000 centrifuges. In 2012, U.S. officials admitted that Stuxnet was the product of U.S. and Israeli experts, and that the attack proceeded as the result of the secret orders issued by President Obama.[4]

Beyond the Infrastructure, there is a significant threat occurring at the Applications layer as well. There are security vulnerabilities in widely used web application software on both the client and server sides. For example, information security training firm SANS issues regular reports on existing vulnerabilities in everything from web browsers, office software, and e-mail clients to Content Management Systems (CMS) and custom-built web application software for businesses. In one analysis, from November 2006 through October 2007, it discovered 4396 security vulnerabilities identified in web application software alone.[5]

Finally, there also exists a threat at the Content layer. Destructive attacks on Web content often occur which either deface websites or take them offline completely. For instance, in August 2011, the Hong Kong stock exchange, the fifth-largest in the world, was the victim of a cyberattack as its web page that publishes important announcements from the market's largest players was taken offline. As a result, trading in the shares of seven companies, including HSBC, Cathay Pacific, China Power International and associated derivatives, had to be suspended. The site stayed offline for a further day, and trading in the dependent positions was also suspended. Another example occurred in 2011 when the hacker collective known as Anonymous, as a show of support for the Occupy Wall Street Movement, launched a coordinated attack that shut down the Oakland Police Department's website the same day that organized protests in Oakland were set to occur, and confidential police data was published on the Web.[6]

These are the main substantive threats that U.S. national cybersecurity policy is designed to address. Our four-layer model not only assists in clarifying the nature of these threats, but also provides insight as to who are the best positioned actors with governing authority at each layer that can create and implement meaningful cyber defenses.

In order to more clearly conceptualize the various types of cybersecurity threats, it is necessary to break down the categorical mechanisms that occur at each layer. Understanding these mechanisms is a process that is again aided by use of our four-layer model: *cyberterrorism* relates primarily to the Infrastructure layer, *cracktivism* and *cyberwarfare* to the Applications layer, and *hacktivism*, being only sometimes threatening, and even then, viewed as far less critical, relating to the Content layer.

Cyberterrorism refers to the use of the Internet in assisting terrorist attacks occurring in the physical world. These attacks are directed at targets in real-space, and thus should be appropriately classified as Internet-enhanced, rather than Internet-based. Cyberterrorism typically targets the Internet's Infrastructure layer and makes use of its costless global communication capacity, coupled with the relative ease of ensuring anonymity and providing difficulty in locating the origin of messages delivered, in facilitating terrorist actions that have outright destruction of their targets as their primary objective.

Examples of how cyberterrorists utilize the Internet for such activities include the recruitment of new members, collaboration with fellow conspirators in planning terrorist actions, and penetrating electronic systems to destroy real-space targets, such as breaking into an air traffic control system in order to make two planes collide, or to shut down emergency 911 services.[7] The Al-Qaeda terrorists who launched the September 11 attacks used the Internet to collect information on flight times, to share information and coordinate their attacks among numerous terrorist cells, and to steal Social Security numbers and obtain fake driver's licenses. The terrorists accomplished much of this by using the Internet in public places and sent messages via public e-mail.[8]

The Internet certainly provides cyberterrorists with several advantages over their counterparts who are not Internet-enabled. Such advantages include a greater chance of ensuring anonymity through the use of proxy servers and IP/subnet masking, and, in fact, software is readily available to assist in providing anonymity to users. The Anonymizer and the Freedom Internet Security package are examples of software products sold to the general public along with a litany of rival packages including Anonymous Surfing, Net Shield, Total Privacy Suite, the Digital Shredder, and Freedom WebSecure.

There are tremendous difficulties for governments seeking to regulate such anonymity software when examined through the lens of our four-layer model. Such software exists at the Applications layer, meaning that virtually any computer programmer can write the code for such anonymity software, and governments' only authority stems from their territorial jurisdiction over the software developers' physical location. However, even when that location is established, governments still encounter difficulties in regulation due to that fact that most of this software is primarily geared toward ensuring privacy from commercial spyware and from the potentially prying eyes of network administrators or ISPs engaging in undesired, and perhaps even unlawful, surveillance—thus the software often has very legitimate and legal uses. Regulation of anonymity software also encounters vast opposition from civil libertarian groups like the EFF who argue that the right to individual privacy cannot be abridged without just cause and after proceeding through the proper legal channels—a process which may often require court orders, subpoenas, and other legal devices, considered vital to preventing unwarranted corporate or government intrusion.

In addition to anonymity, cyberterrorists have the added advantage of global instantaneous reach; and again, determining the physical locations where messages originate makes policing cyberterrorist communications highly problematic.[9] The Internet as a global communications medium enables the costless mass dissemination of such communications, and the forums being utilized, even when discovered and shut down, are typically mirrored on servers that reside in multiple countries around the world. Mirrored forums create a whack-a-mole regulatory environment where, even when law enforcement succeeds in discovery, the total shutdown of a cyberterrorist communications network is virtually impossible because immediately after a forum is taken offline, users simply redirect to one of its mirrored sites.

Cyberterrorists also have an advantage in their use of steganography—the practice of disguising messages within digital images. For instance, a digital image of a sailboat might contain a micro-image of a map that would only be visible if someone knew to look for it, or a digital song file might contain blueprints of a targeted building.[10] Such steganographical measures not only render the secretive content of messages largely undetectable by data mining software filters, but also are easily accessible through common commercial software applications—meaning that having a working knowledge of computer programming is not a prerequisite.

Meanwhile, *hacktivism* refers to computer hacking for political purposes. Narrowly defined, this refers to using programming and inter-networking technologies to disrupt the electronic activities of its targets which exist primarily in cyberspace. As a result, it is most appropriately classified as occurring at the Internet's Application and Content layers.

Hacktivism is based on the original meaning of the term "hacking," made famous during the 1970s in the context of telephony, which is to create a clever or quick solution to a problem which works outside of the generally accepted norms of the environment. To hack something is not necessarily nefarious; it is simply to use a technology for a purpose other than for which it was originally intended. The common use of the term in popular parlance, such as "My computer was hacked and my credit card number was stolen," is mislabeled as hacktivism. As will soon be explained, the destructive act of breaking into a computer or network for criminal purposes is actually computer cracking, not hacking. Hacktivism does not actually seek to steal from or destroy its cybertarget. In a political context, disruption, not destruction, is its primary motive with its ultimate goal being to attract attention to specific issues.

For example, hacktivists have launched "Google bombs" in every presidential election cycle since 2004. This refers to computer programmers gaming Google's search algorithm so that, for instance, when a user searched for the phrase "miserable failure," the top result displayed a link to President George W. Bush. Hacktivists were able to accomplish this, not by cracking into Google's servers and stealing or destroying content, but simply by learning how Google's search algorithm functioned—ranking its results in a certain order based on the number of external links to a website. As a result, hacktivists were able to recruit enough links to selected pages about President Bush, all incorporating the phrase "miserable failure," so that once enough external links were created, Google's algorithm ranked it accordingly and directed users to the biography of the president on the official White House website.[11]

Other examples of Google bombs include when, in 2008, a search for either "failure" or "cheerful achievement" both produced results to Barack Obama,[12] while a search for "John McCain" produced stories about the Senator's filibustering of a minimum-wage hike.[13] Also, in 2012, another Google bomb was launched directing searches for "completely wrong" to multiple photos of Mitt Romney.[14]

Using the same logic, "Twitter bombs" have developed more recently where hacktivists seek to hijack the hashtags of their political adversaries.

For example, in 2008, a Republican-leaning organization calling itself the "Don't Go Movement," whose mission was to persuade Congress to stay in Washington D.C., and not go on summer recess, until a solution for the U.S. energy crisis was found, called on Twitter supporters to include the "#DontGo" hashtag in all of their posts. Overnight, "#DontGo" became the top trending topic on the entire site. However, not to be outdone, Democratic-leaning hacktivists, dismayed at #DontGo's success, began an organized effort to pollute the Twitter stream. In other words, critics of the #DontGo Movement were being encouraged to also include the #DontGo hashtag in their own posts, that way when people searched Twitter, more critical posts would be displayed.[15]

Another example of hacktivism is an "edit war" that occurs between competing parties when they publicly use open comment-based systems to disseminate their views. One such example of an edit war occurred in 2008 when members of the group self-labeled "Anonymous" launched a campaign to write reviews of L. Ron Hubbard's book, "Dianetics"—the foundation of the Church of Scientology—on its Amazon.com website. Most of the reviews were decidedly negative and many did not even address the book, but rather criticized the Scientology movement in general. Scientologists responded by using the same comment-space to rebuke those criticisms, and this back-and-forth edit war ensued until Amazon was forced to remove nearly all of the comments for an indefinite amount of time.[16]

There was also the edit war that occurred on Wikipedia over its entry on the 1948 Arab-Israeli War. A pro-Israel group, the Committee for Accuracy in Middle East Reporting in America (CAMERA) publicly called for volunteers to edit Wikipedia entries that displayed "notable bias" on the subjects of the 1948 war and Israeli Independence. In response, a Palestinian aggregator called the Electronic Intifada exposed the initiative, encouraging its supporters to also volunteer as editors to "ensure that these articles are free of bias and error." The result was that the published entries displayed very different information based solely on who was the last editor to revise them.[17]

Additional examples of hacktivism include the reporting of liberal or conservative blogs to various ISPs and website operators as spam; also, the EFF's development of its Tor network, and the setting up of proxy servers in general, to remotely assist democratic activists in China or Iran.

There are other, more classic examples of hacktivism with relatively established and long-standing histories. These include various forms of electronic civil disobedience, notably website black-outs and virtual sit-ins.

One of the earliest cases of hacktivism was the "Turn the Web Black" protest, also termed "Black Thursday," which occurred on February 1, 1996 when a large number of websites changed their background color to black for 48 hours in an effort to raise public awareness about the Communications

Decency Act, and to what participants argued was the Act's infringement on free speech.[18] An electronic alert was circulated via e-mail distribution lists and electronic discussion boards to spread by word-of-mouth the plan for this blackout. While certainly the group known as the Voters Telecommunications Watch played a crucial initial role in formulating the planned action and drafting the electronic alert, what occurred thereafter was a decentralized method of bringing about collective action, achieved at the grassroots level. This might appropriately be described as a digital viral marketing strategy applied to protest coordination. Thousands of websites participated, and the event captured the attention of such traditional news media outlets as the New York Times and CNN.[19]

The case of the "Turn the Web Black" protest also holds immense significance in that it represents hacktivism where the principle recruitment target was not the mass public, but rather a narrow base of computer programmers and website operators. Confirming our political architecture at the Applications layer, this was an acknowledgment that only website operators would have, first, the authority to alter the design of their websites, and second, the technological capability to do so. Thus, the protest's call for participants was not necessarily directed at all individuals making up the mass public to contribute, but rather only to the select group of website operators in control of the code behind websites—a relatively small percentage of the public, particularly in 1996.

A more recent example of hacktivism involved the website Wikileaks. In November 2010, approximately 250,000 classified documents from the U.S. State Department were posted on Wikileaks—a website self-described to be a safe haven for whistle-blowers with its stated purpose being to expose corruption in both the government and the private sector.[20] Immediately upon these documents being released, it became clear that the individual who leaked them committed a criminal act according to U.S. law, however what was less clear was whether Wikileaks, the website, had actually broken the law, playing no role other than hosting the materials, acting solely as a forum where someone else posted content—a characterization which U.S. courts have often recognized as within the legal realm.[21] Wikileaks, in this way, is an example of a hacktivist website, pursuing political aims (the exposition of corruption) by utilizing Internet technology in a novel way—its legal status being hazy, at best, and disruption being the primary motivator rather than destruction.

Electronic civil disobedience has occurred as both a stand-alone activity and also a joint venture with real-space protests. Stand-alone hacktivism occurs when a collection of individuals in cyberspace perpetuate some action against a target which also resides in cyberspace. An early prominent example of such stand-alone hacktivism would be the virtual sit-in organized by the Electronic Disturbance Theater (EDT) in support of the Zapatista movement in Chiapas against the Mexican government.[22] In order to show their opposition to the actions of the Mexican government, the EDT created a virtual sit-in software tool called FloodNet, whose purpose was to temporarily prevent Internet users from accessing the Mexican government's website by means of flooding—in this case, directing thousands of "hits" by protest participants at the website until the web server would be unable to handle any more.

While certainly the Zapatista rebels exist in real-space, it is of great significance that the EDT acted independently and did not consult with either the Mexican government or the Zapatistas themselves. The EDT's virtual sit-in was a cyberaction directed exclusively at a cybertarget. In fact, if you had no Internet access, then there would have been no way to ascertain that a protest even occurred.

Electronic civil disobedience has also taken the form of being part of a joint venture with real-space protests—in other words, hacktivism coordinated in conjunction with real-space activism. Perhaps the most famous early example is the 1999 virtual sit-in that coincided with the World Trade Organization (WTO) protests in Seattle. Organized by the Electrohippies Collective, it was intended to flood the WTO's web servers simultaneously with the massive street protests.[23] This joint-venture style of hacktivism, where cyberaction is only one element in a larger, more coordinated protest strategy, is designed to allow supporters of a cause who physically cannot attend a street-based protest to still contribute and make their presence felt through alternative means. In this sense, hacktivism enables mass mobilization and collective action, overcoming the traditional barrier of geography.

While this joint-venture approach to hacktivism contextualizes it as having a noticeable impact toward a larger social protest, it fails to adequately address hacktivism *as* the social movement. Its proponents argue that technology and the Internet are more than simply new tools for protest movements; they are fundamentally a new type of movement. Contrary to a cyberactivist model, whereby the Internet is simply a means of better organizing collective actions through interactive communication technologies, the true power of hacktivism lies in the fact that, in cyberspace, mass mobilization need not even be vitally necessary for the furthering of protest movement objectives, so long as electronic disruption can be achieved through programmed automation.

That aside, again, the overriding goal of hacktivism is to temporarily disrupt—not destroy—the digital activities of its target in order to disseminate one's views and bring attention to an issue. It is ultimately a form of protest, and, as such, hacktivism is not a primary threat to national cybersecurity, and, as we shall see, the *National Strategy* policy document bears that out, except in cases when the hacktivist is disrupting online activities that are deemed "critical" to national security or in a manner deemed threatening enough to warrant a strong defense.

While hacktivism is not necessarily a primary threat to national cybersecurity, cracktivism most certainly is. *Cracktivism* refers to activities which seek the willful destruction of cybertargets. This can be any unauthorized intrusion, defacement, or act of causing intentional damage using internetworked technologies and other electronic means. Furthermore, cracktivist activities are generally illegal. Many acts of cracktivism are often mistaken as hacktivism, such as the releasing of viruses or worms, however these actions are not so much seeking to further the political goals of a broader movement as they are to destroying their targets altogether. Cracktivism is clearly one of the largest threats that national cybersecurity policy is meant to address.

The most common forms of cracktivism come in the form of computer viruses, worms, Trojan horses and other malicious code attacks. It is through computer cracking that some also aim to break into secure systems with the intent of stealing vital information or outright destroying it, and these unauthorized intrusions are particularly problematic in an Internet environment where attacks can more easily be carried out remotely.

Cracktivism was evident in the 2006 uproar over cartoons published in a Danish newspaper depicting the Prophet Mohammed. As mass protests in the Arab world took place over the cartoons, a "cyber-jihad" simultaneously occurred, whereby more than 1,000 Danish, Israeli, and European websites were defaced or completely shut down by Islamic cracktivists.[24] In retaliation, several pro-Western and pro-American cracktivist groups, such as the Freedom Cyber Force Militia, defaced prominent Arabic websites including AlJazeera, hijacking their domain name and redirecting users to different servers.[25] This quickly escalated into a cyberwar between cracktivists on both sides.

The ongoing cyberwar between Israeli and Arab programmers also holds many examples of cracktivism. Throughout the conflict, both sides have engaged in willfully destructive activities, notably website defacements, e-mail bombs, and ping storms. For example, an Israeli teenage cracktivist successfully spread a virus through an Iraqi government website during the reign of Saddam Hussein, locating the web server with the assistance of special software tools, and then including a virus in an e-mail attachment for Iraqi officials to open.[26] Additionally, pro-Palestinian attack sites began publicly distributing the Melissa and LoveLetter viruses, as well as various Word macro viruses, for use against Israeli sites in 2001. This marked one of the first confirmed uses of distributed viruses being part of a larger coordinated attack strategy used in a cyberconflict.[27]

Another example of cracktivism is the case where three teenagers broke into the computer systems of India's Bhabha Atomic Research Centre (BARC) in Bombay in response to the country's nuclear weapons tests. They defaced the research center's home page, claimed to have stolen e-mail messages exchanged by the nuclear scientists from the center's servers, and actually erased data pertaining to the nation's nuclear weapons programs.[28]

To be clear, this type of activity is not hacktivism, properly defined. When dealing with disrupting the activity of a cybertarget through electronic means, the fundamental difference must again be stated—that hacktivism seeks to disrupt its targets' activities in order to bring attention to the political goals of its broader movement, affecting changes in policy, whether public or private. Cracktivism, however, while sometimes motivated by political goals, seeks to destroy its cybertarget and willfully cause damage through illicit means. As a result, acts of cracktivism often prove counterproductive (assuming there was a true political motive in the first place) insofar as the targets respond by denouncing the actions and actors involved as illegitimate, soliciting a decidedly negative reaction from observers, and drawing attention away from any political goals of the cracktivist and instead toward the illegal action itself.

Finally, *cyberwarfare* refers to the deployment of digital attack technologies by one governmental entity against another, or against an enemy of that government. Often, cracktivism makes up the major component in cyberwarfare strategies, however, rather than the aforementioned examples of cracktivism being on behalf of one group or individual against another, cyberwarfare involves similar attacks but on behalf of one government against another, often implemented by formal military units. Clearly, this is another primary threat that U.S. national cybersecurity policy is designed to address.

Cyberwarfare incidents, like cases of cracktivism, occur mostly at the Applications and Content layers. Such incidents have strikingly risen and become more prominent over the past decade. There was a cyberwar between Russia and Estonia in 2007 where the Estonian authorities removed a statue of a World War II-era Soviet soldier, prompting ethnic Russians—and allegedly the Russian government itself—to attack Estonian cybertargets including nearly "shutting down the country's digital infrastructure, clogging the Web sites of the president, the prime minister, Parliament and other government agencies, staggering Estonia's biggest bank and overwhelming the sites of several daily newspapers."[29] The Russians did all of this, not by "hacking" into Estonian computer systems the way the mainstream public often misuses the term, but by launching DDOS attacks. Basically, software bots turned computers around the world into "zombies" that sent and requested so much data from Estonian servers that they became overloaded and shut down from not being able to handle all of the traffic.

Another cyberwar is ongoing between the U.S. and Al-Qaeda. The New York Times reported in 2008 that "to counter efforts by terrorists to plot attacks, raise money and recruit new members on the Internet, the government has mounted a secret campaign to plant bogus e-mail messages and website postings, with the intent to sow confusion, dissent and distrust among militant organizations, officials confirm."[30]

Meanwhile, there is much evidence to suggest that China is actively engaged in a cyberwar with the U.S. as well. In 2008, the Defense Department cited numerous successful attacks in the previous year originating in China, including shutting down Homeland Security networks and hacking into the Pentagon e-mail system used by the offices of Defense Secretary Robert Gates.[31] Additionally, in December 2009, reports surfaced that a sophisticated cyberattack against Google and 30 other U.S. companies was traced back to "a single foreign entity consisting either of agents of the Chinese state or proxies thereof."[32] More recently, in February 2013, evidence was gathered linking an "overwhelming percentage" of attacks on American corporations, organizations, and government agencies from originating within a single 12-story office tower on Datong Road off the outskirts of Shanghai—which is the headquarters of the People's Liberation Army's Unit 61398.[33]

There is a related security challenge that cyberwarfare highlights: that a continual build-up in ever-more effective cyberattacks and cyberdefensive countermeasures inevitably leads to, what can best be conceptualized as, a cyber arms race. Escalating cyber arms races, occurring at the Applications layer, shift the balance of power heavily toward computer programmers, as the cyberwar becomes truly a conflict between competing software tools and programming expertise on the two sides.

In response to the acknowledged threats posed by the digital arms races stemming from cyberwarfare, the U.S. and Russian governments have held discussions on mitigating potential effects. In 2009, it was reported that Russia favored an international treaty along the lines of those negotiated for chemical weapons, essentially looking to ban offensive weapons and tactics. On the other hand, the U.S. instead advocated for improved cooperation among international law enforcement groups, basically trying to formalize the criminalization of such acts through legal channels.[34]

In summary, there are three primary threats which U.S national cybersecurity policy is designed to address: cyberterrorism, cracktivism, and cyberwarfare—with a fourth, hacktivism, only sometimes threatening, and even then, viewed as far less critical. As with all typologies, these categories of cybersecurity threats are not always mutually exclusive. Cases will inevitably arise with certain characteristics which may fall into more than one category or none at all. Denial-of-Service attacks, for example, are sometimes relatively harmless cases of hacktivism, and sometimes immensely damaging cases of cracktivism—depending on the level of disruption or destruction they intend to produce. This typology is, therefore, not intended to be mutually exclusive, but rather to serve as a general categorization of the cyber threats that national policy is intended to defend against, in an attempt to conceptualize the cybersecurity problem stream within the context of our four-layer model.

With that typology established, we can further break down specific cybersecurity threats according to their particular attack mechanisms. Cyberterrorists, cracktivists, and agents of cyberwarfare all seek to achieve their objectives through three primary means: *Malicious Code, Distributed Denial-of-Service Attacks, and Unauthorized Access*. National cybersecurity policy is geared toward defending critical assets against these potentially destructive forces.

Malicious code attacks refer to viruses, worms, Trojan horses, or other code-based malicious entities that infect a host. This is a threat almost completely occurring at the two code layers of the Internet—the Protocols and Applications layers. As such, prevention and response policies are also heavily geared toward technical solutions.

Viruses are the best known type of malicious code attack, and they come in many forms: file infector viruses, boot sector viruses, and macro viruses. All are designed to infect a host computer by inserting itself into another program with the intention to "destroy data, run destructive or intrusive programs, or otherwise compromise the security or the confidentiality, integrity, and availability of the victim's data, applications, or operating system." All of this is generally executed without the system user's knowledge.[35]

Worms are another form of malicious code attack. Similar to viruses, they are self-replicating and self-propagating, with the major difference being that they do not require a host program to infect a victim, but rather they are self-contained. They can execute themselves without the trigger of a user intervention.

Trojan horses are non-self-replicating programs that appear to be benign but actually have a hidden malicious purpose like replacing existing files or else adding new applications without altering existing files. Because of their benign appearance, Trojan horses are often very difficult to detect.

Mobile code attacks are those that are transmitted from a remote system to be executed on a local system, and have become popularized because web browsers and e-mail clients often grant default privileges to legitimate mobile code applications, thus making their exploitation fairly simple. Mobile code attacks can be carried out through such benign programming languages as Java, ActiveX, JavaScript, and VBscript—all of which are common and accepted technologies permitted on the most pervasive web browsers.

Botnets refer to a collection of compromised computers (also known as "zombie computers") running software that a botnet's originator (a.k.a.—the "bot herder" or "bot master") can control remotely, usually for nefarious purposes. These botnet armies can lie dormant for extended periods of time, constantly growing in number, until programmed to carry out an action simultaneously and with all machines working in conjunction with one another.

Denial-of-Service (DoS) attacks refer to remote attempts to prevent or impair the use of networks, systems, or applications by exhausting computing resources. Reflector attacks, Amplifier attacks, and Synfloods are all types of DOS attacks. The central idea is that by sending too much traffic to a specific website or other target, their resources will be overwhelmed and thus become unavailable, if not shut down entirely. DOS attacks have become increasingly severe, threatening significant network disruptions and major financial loss. No organization can completely protect itself from DOS attacks insofar as they can't necessarily limit how many people attempt to visit their public website (nor do they typically want to).

Finally, unauthorized access is what occurs when a person gains logical or physical access without permission to a network, system, application, data, or other resource. It is usually acquired through exploiting software code vulnerabilities, obtaining passwords, or social engineering. Different from the other threat types, attackers will typically engage in multiple stages of reconnaissance activities in order to map networks, identify hosts, determine what software is running, and discover what vulnerabilities exist; and only then will the most damaging actions be attempted.

Thus, defining the problem which national cybersecurity policy is meant to address can be broken down accordingly. The single generalized problem is protecting the nation's critical cyber assets from attack. However, as we have demonstrated, that broad definition encompasses numerous categories of attacks, often with different goals underlying them, and occurring at different

Table 7.1 The Cybersecurity Policy Stream

Layer-specific Threats	*Categorical Mechanisms*	*Specific Mechanisms*

Infrastructure
• Destruction of physical infrastructure; Hijacking of industrial control systems
 	
Applications
• Stolen data; Network devices being disabled or hijacked
 	
Content
• Defacement of websites; Websites being taken offline completely
	Cyberterrorism
	Hacktivism
	Cracktivism
	Cyberwarfare
		Malicious Code
		• Viruses
		  ◦ File infector viruses
		  ◦ Boot sector viruses
		  ◦ Macro viruses
		• Worms
		• Trojan horses
		• Mobile code
		• Botnets
		
		Denial-of-Service (DoS) Attacks
		• Reflector attacks
		• Amplifier attacks
		• Synfloods
		
		Unauthorized Access

layers. Hence, we have developed a typology whereby the categorical threats of cyberterrorism, hacktivism, cracktivism, and cyberwarfare are all made distinct, and their corresponding specific attack mechanisms are placed in this context. With the problem stream established, it is now possible to more effectively analyze the actual policy at the heart of the U.S. cybersecurity—the Bush administration's *NSSC*.

THE POLICYMAKING PROCESS AND POLICY DESIGN BEHIND THE NATIONAL STRATEGY TO SECURE CYBERSPACE (NSSC)

- Pre-9/11 History
- Executive Order creating the CIPB
- 53 Questions
- Public Comments
- Draft version of the Policy
- Town Hall Meetings
- Final version of the Policy

In the aftermath of the terrorist attacks of September 11th, 2001, President George W. Bush initiated the process for formulating what would eventually become known as the *NSSC*, the most comprehensive and ambitious federal policy to date in addressing the challenge of protecting the nation's critical cyber assets from attack.

The policy was ambitious, and considered by the Administration to be an expansive evolution of "the first attempt by any national government to design a way to protect its cyberspace." The policymaking process can be characterized as open, but flawed. A presidential advisory board released 53 questions to the public for comment, then drafted an initial proposal which was discussed in several town hall meetings across the country, ultimately leading to the final version of the policy—the *NSSC* document.

This policymaking process was heavily influenced at every stage by large private corporations and, from the outset, its policy design came under heavy criticism for relying on a strictly voluntary public-private approach. We will argue that its implementation was greatly hindered by the federal government's failure to allocate enough resources to the problem, by a high turnover rate of top officials within the newly created Executive bureaucracy in its early years, and by organizational conflict between competing federal agencies in subsequent years. Ultimately, by analyzing the policy through the lens of our four-layer model, we will argue that its central problem lies, not with its policy design, but with its implementation.

The following is the story behind the *NSSC*.

As we previously detailed in our historical narrative, while the Internet dates back to the 1950s as a Cold War product of the U.S. Defense Department, it wasn't until the popularization of the World Wide Web in 1993 that scholars mark a turning point in the Internet's significance to the global culture and economy. Similarly, while U.S. cybersecurity policy has origins that span several decades, it wasn't until the focusing event of the terrorist attacks of September 11th, 2001 that the issue of a national cybersecurity policy proportionally grew in relevance and visibility.

Certainly, it is instructive to briefly examine the federal government's history of taking protective cybersecurity measures that precede the main focusing event of the September 11th, 2001 terrorist attacks, for it illustrates the policy context in which the *National Strategy* ultimately came about; for it certainly was not wrought in a vacuum.

In the aftermath of the 1995 bombing of the Alfred P. Murrah Federal Building in Oklahoma City, President Bill Clinton, having heightened concerns over the security of other important facilities and national landmarks, established the President's Commission on Critical Infrastructure Protection (PCCIP) in 1996. The Commission then surprised many observers by generating a report titled, "Critical Foundations: Protecting America's Infrastructures,"[36] that "did not focus on the vulnerability of key buildings around the country but instead on the security problems in the new phenomenon of cyberspace."[37]

The PCCIP would lead to Presidential Decision Directive 63 (PDD-63) in May 1998, which stressed the fundamental importance of cooperation between the government and the private sector, explicitly noting that this is necessary because "nearly 90 percent of [the nation's vital information networks] are privately owned and operated."[38] This is significant for two reasons—first, it signals the earliest indicator of the federal government's acknowledgment that cybersecurity policy be necessarily focused heavily on private-sector cooperation, rather than direct regulation, due to the Internet's decentralization of governing authority at different layers; and second, the PCCIP would later evolve into the Bush administration's President's Critical Infrastructure Protection Board (PCIPB) which would directly author the *NSSC* policy.

In 1997, the Pentagon ran an exercise called Eligible Receiver under joint White House and Defense Department supervision with the purpose of attempting to probe the Defense Department's networks from the Internet, "without the benefit of any insider knowledge." The team running the exercise wanted to prove that even the Pentagon was vulnerable to attack by gaining unauthorized access to their network, only within two days they gained access all the way into the Joint Chiefs' command system and they could alter messages going out from the Pentagon. The exercise was stopped immediately and military departments were ordered to procure and install new intrusion detection systems to detect possible intrusions into their networks that were not authorized. They immediately showed that "thousands of attempted illegal penetrations of DoD networks were going on every day."[39]

In February 1998, amid fears that the "Y2K bug" could potentially cripple cyber systems at the start of the new millennium, President Clinton appointed former Deputy Budget Director John Koskinen to chair his Year 2000 Conversion Council. This council centralized executive branch efforts to prepare government agencies for the Y2K date rollover, but perhaps more significantly, the council also became the template for later Executive branch efforts to centralize oversight of cybersecurity threats.[40]

In a case that the Federal Bureau of Investigations (FBI) named Solar Sunrise, the logistics systems at many Air Force bases were penetrated during the same weekend that President Clinton ordered a military deployment to the Arab Gulf in response to Iraq's refusal to allow U.N. weapon inspectors. While suspicions circulated that Iraq was engaging in cyberwarfare, the unauthorized activity was actually traced by the FBI, with the help of several private software firms, to three teenagers—two located in California, the other in Israel.[41]

In March 1998, another incident known as Moonlight Maze began whereby a pattern arose of probing computer systems at the Pentagon, NASA, the Energy Department, private universities, and research labs. Access had been gained to thousands of files including "maps of military installations, troop configurations and military hardware designs" over the course of two years. The Defense Department traced the intrusions to a mainframe computer located within Russia, "but the sponsor of the attacks is unknown and Russia denie[d] any involvement."[42]

In May 1998, President Clinton also appointed Richard A. Clarke, a former staffer at the National Security Council (NSC), as National Coordinator for Security, Infrastructure Protection and Counter-terrorism.[43] Clarke would remain a top counterterrorism and cybersecurity advisor to both President Clinton and President Bush until January 2003, and would be the primary individual responsible for drafting the *NSSC* policy.

In 2000, after the world transitioned through Y2K with no major disasters, Rep. Stephen Horn (R-CA), chairman of a House Government Reform subcommittee, officially changed the "Y2K Readiness Report Cards" he had been publishing quarterly for federal agencies since 1996 into "Cybersecurity Readiness Report Cards." Many federal agencies continued to receive failing grades. After the Clinton administration dismantled the Y2K Center, some members of Congress began calling for the appointment of a federal chief information officer—or "cybersecurity czar"—to oversee privacy and security issues.[44]

Also, in January 2000, the Clinton administration released a new national cybersecurity strategy. Titled, *The National Plan for Information Systems Protection*, this proposal focused on the federal government serving as a model of information security and for building public-private partnerships.[45] The document, although billed by the Administration as "the first attempt by any national government to design a way to protect its cyberspace," earned a "cool reception" from the private sector, which was left out of much of the drafting process.[46]

The remainder of the year 2000 witnessed numerous DDoS attacks and the outbreak of the "I Love You" computer virus, which wrought havoc on government and commercial systems worldwide, even bringing down several of the world's largest and most popular portal and e-commerce websites. Congressional hearings and legislative proposals immediately followed aimed at enhancing the nation's cybersecurity—both from public- and private-sector perspectives.[47]

While the need for an effective national cybersecurity policy was already finding its way gradually onto the national political agenda, all of these events would soon be overshadowed by the central focusing event in American cybersecurity policy: the terrorist attacks of September 11, 2001.

In the aftermath of the 9/11 attacks, President George W. Bush delivered a speech to a joint-session of Congress that called for the creation of a new Cabinet-level agency within the federal government—the DHS. The president charged the new Department with the responsibility of providing "the unifying core for the vast national network of organizations and institutions involved in efforts to secure our nation."[48]

Within months of its creation, DHS, through Executive Order 13231 signed by the president, set up the Critical Infrastructure Protection Board (CIPB), headed by the president's new top cybersecurity advisory Richard A. Clarke, in order to create a draft version of a national cybersecurity policy.[49] The CIPB would be composed of senior officials from more than 20 departments and agencies. Original plans called for the final version to be released on September 19, 2002, complete with a presidential signing ceremony at Stanford University amid technology icons like Microsoft chairman Bill Gates, however the White House decided to hold back the final plan and push back the release date.

Instead, the CIPB decided to release the draft version of the strategy for a five-month period of public comment, and ten town hall meetings were held around the nation to gather further input.

This period of public comment was initiated by the CIPB issuing a document titled, "53 Questions for Developing the National Strategy to Secure

Cyberspace."[50] The 53 questions were intended to initiate discussion and act as a jumping-off point for public comments to be submitted, but they also serve to illustrate the Bush administration's early policy preferences, which would later appear in the final draft of the NSSC.

For example, the 53 questions are explicitly divided into 5 levels, or categories: "the home user and small business," "major enterprises," "sectors of the national information infrastructure," "national level institutions and policies," and "global." While each of these categories is provided with a range of issues that ought to be addressed by the NSSC, the number of questions raised within each category clearly demonstrates the Administration's central focus. The home user and small businesses are provided with a total of 5 questions, while major enterprises received 11, sectors of the NII received 24, national-level institutions and policies received 14, and, finally, the "global" category was addressed with a total of 2 questions.

These numbers are clear evidence of the Administration's perception of the cybersecurity dilemma, from the earliest stages of the policy process, as being one to be handled primarily by major industries in the private sector and federal-level institutions.

Furthermore, the framing of the questions also reveals certain preconceived biases as to how cybersecurity solutions ought to be formulated. For example, there are numerous references to using market forces as an alternative to governmental regulation, as well as a repeated emphasis on the private sector establishing "standards and best practices," and engaging in self-evaluation, training, reporting, information-sharing, and outsourcing as a means for safeguarding systems. By contrast, only a relatively small number of questions relate to governmental regulation or the creation of mandates, and those that do only apply to the government's own systems, not those of the private sector or home user. In other words, the language of the 53 questions direct the reader, and, ultimately, the public comments, in certain directions from the earliest stage of the policy process.

Another noteworthy element of the 53 questions is that all of the four conceptual layers are at least minimally addressed. The Infrastructure layer is clearly the focus of the document, however there are also questions related to the Protocol layer,—the 802.11 wireless standard, for example,—the distribution of software patches, which go directly to the heart of the Applications layer, and finally, questions about adjusting liability laws and criminal justice penalties which address the Content layer. However intentional or unintentional this may be, the fact that all four of the conceptual layers are at least somewhat addressed highlights an implicit understanding of their importance in formulating comprehensive Internet public policies.

The CIPB also released a preliminary draft version of the *National Strategy* in September 2002. As with the 53 questions, this draft version was intended to act as a jumping-off point and to stimulate public comments. It was signed by the Chair and Vice-Chair of the CIPB, Richard A. Clarke and Howard A. Schmidt.[51]

The draft version includes many of the elements and key strategies that would make it into the final version of the NSSC, such as recognizing explicitly "the reality that the Federal government alone cannot secure cyberspace" and that cybersecurity "depends on a public-private partnership" where "everyone must act to secure their own parts of cyberspace."

The guiding policy principles are listed in the draft version as follows: (1) embrace public-private partnerships, (2) avoid regulation, (3) safeguard civil liberties and privacy, (4) coordinate with Congress, and (5) cooperate with State and local governments. Out of the over 20 agencies included in Executive Order 13231, the following are designated as "lead agencies" for the protection of critical infrastructure: the DHS, Treasury, Health and Human Services, Energy, the Environmental Protection Agency, Agriculture, and Defense.

Thus, the draft version clearly reveals the Administration's guiding policy design principles, making them far more explicit in this draft version than in the 53 questions, where those principles were also evident, though unstated.

Additionally, as was also the case with the 53 questions, a breakdown of its content helps define the Administration's central focus. The draft version uses the same "5 levels" or categories of issues that ought to be addressed. Within each level, the draft calls for specific recommendations, programs, and discussions, which are suggested in the following quantities.

These numbers indicate that the draft version of the NSSC places a clear emphasis on major sectors and federal-level institutions, with relatively little focus on the individual home user or on adopting a global approach.

The main purpose of both the 53 questions and the draft version of the NSSC was to stimulate discussion and the submission of public comments to gather input for what would ultimately lead to the final NSSC policy. After being drafted, both were made accessible by being placed on web pages sponsored by government agencies, associations, and private organizations.

Table 7.2 Quantitative Breakdown of the NSSC Draft Version's Recommendations

		Recommendations	*Programs*	*Discussions*
Categories	Home User & Small Business	5	9	2
	Major Enterprises	7	5	4
	Sectors of the National Information Infrastructure	19	15	9
	National-Level Institutions & Policies	49	0	28
	Global	6	6	1

The public comments themselves were submitted primarily by insiders within the technology industry. Michael Rasmussen—V.P. for Standards and Public Policy at the Information Systems Security Association—would later issue this description of DHS summits:

> You have a lot of IT vendors lobbying Capital Hill trying to convince legislatures that security is completely technical and what we need is more products . . . DHS and legislators need to get more input from the people in the trenches. The summit did reach out to many, but it was organized by the high tech sector. I would have liked to have seen more end-user organization involvement. Particularly Chief Information Security Officers or Chief Risk Officers."[52]

Likewise, another common criticism of the public commentary process was that business lobbyists had undue influence, embodied in statements such as the following one from the senior editor of About.com: "Clarke's task force quickly ran into opposition, mostly from wealthy lobbyists representing communications, software, and security companies, but also from (surprise!) the White House."[53]

The town hall meetings, sponsored by the White House, were held in 10 metropolitan areas, and sought to solicit views from both the public and private sectors. Individual sectors, such as higher education, state and local government, banking and finance, etc., formed workgroups to create initial sector-specific cyberspace security strategies. The town hall meetings occurred in Denver, CO, Portland, OR, Chicago, IL, Atlanta, GA, and several other cities nationwide.

These town hall meetings often featured Richard Clarke as the principle speaker, as was the case in San Diego, Denver, Washington D.C., and Portland. In other cases, Howard Schmidt (the CIPB Vice-Chair) was the principle speaker, such as at the University of Pennsylvania.

Notifications of the town hall meetings were often sent out via e-mail distribution lists, and a public website was also established to disseminate the details of the meetings.

In addition, the Commerce Department's Critical Infrastructure Assurance Office (CIAO) sponsored meetings with state and local government officials from several states, which included national-level conferences held in Austin, Texas and Princeton, New Jersey in February 2002 and April 2002, respectively.[54]

It was at this time that the White House renamed the presidential advisory panel formerly known as the CIPB as the National Infrastructure Advisory Council (NIAC) which consisted of "leaders from the key sectors of the economy, government, and academia." The president's National Security Telecommunications Advisory Committee (NSTAC) also reviewed and commented on the draft.[55]

Ultimately, after a year of research by businesses, universities, and government officials, on February 14, 2003 the DHS unceremoniously released the final version of the draft, titled *The National Strategy to Secure Cyberspace (NSSC).*

In the Political Science literature, policy designs are characterized as significant insofar as they create meaningful consequences for democracy.[56] They are intentional and purposive creations,[57] yet, once the guiding principles are put in place, they become more acutely defined by accidents, external forces, and a highly iterative bureaucratic process.[58] Context is considered the single most important predictor of what type of design will result. Policy designs are crafted in ways that are tailored to fit some conception of the situation, sometimes reflect multiple and conflicting values, and typically have consequences for that context from which they emerged.[59] As is the case with the policy design of the NSSC, decision-making is an ongoing contextual process determined by events "on the ground" as much as it is by one-off rational selections between alternatives occurring early in the policy process.

The NSSC is an Executive report and a component of the larger national strategy for homeland security. Its policy design calls for government recommendations—not mandates—to be issued to businesses, individuals, and government agencies to secure their own respective computer systems and private networks. It relies on voluntary actions and public-private partnerships for implementation.

The NSSC's stated purpose is "to engage and empower Americans to secure the portions of cyberspace that they own, operate, control, or with which they interact." The NSSC has three strategic objectives: first, to prevent cyber attacks against America's critical infrastructures; second, to reduce national vulnerability to cyber attacks; and third, to minimize damage and recovery time from cyber attacks that do occur.

Furthermore, the document lays out its six guiding principles: (1) to make cybersecurity a national effort; (2) to protect the privacy and civil liberties of "consumers and operators"; (3) that market forces, rather than direct federal regulation, are expected to provide the major impetus of cybersecurity; (4) to bring these about through the assignment of responsibilities and accountability to federal, state, and local government agencies, as well as the private sector; (5) to ensure flexibility in the government's ability to respond to cyber attacks and manage vulnerability reduction; and (6) to adopt multiyear plans for sustaining cybersecurity into the future.[60]

The government's role in cybersecurity is made explicit. Implicitly acknowledging the political architecture of governing authority that our four-layer model produced, the policy states that:

> "in general, the private sector is best equipped and structured to respond to an evolving cyber threat . . . externally, a government role in cyberspace is warranted in cases where high transaction costs or legal barriers lead to significant coordination problems, cases in which governments operate in the absence of private sector forces, resolution of incentive problems that lead to under provisioning of critical shared resources, and raising awareness."

As a consequence, the NSSC focuses on public-private partnerships in order to carry out its five national priorities. The first priority of the NSSC is to create a National Cyberspace Security Response System which would perform analyses, issue warnings, and coordinate response efforts. The plan emphasizes the accomplishment of these objectives through "encouraging the development of private sector capability," and such a response system would be implemented through partnerships between government and industry at the local, state, and federal levels separately to ensure the "health" of cyberspace.

The second priority of the NSSC is to create a National Cyberspace Security Threat and Vulnerability Reduction program to address weaknesses in technology and the improper implementation and oversight of technological products. The power of law enforcement agencies would be enhanced and a process would be created for vulnerability assessments. Additionally, the "mechanisms" of the Internet are to be secured by improving protocols and routing, improving the security of the physical telecommunications infrastructure, and by fostering the use of new digital control systems and data acquisition systems—effectively giving more "supervisory control" to law enforcement agencies over the Internet activities of users. This is a clear recognition of the importance of both the Infrastructure and Protocol layers as they relate to cybersecurity.

The third critical priority is to create a National Cyberspace Security Awareness and Training Program to address security education for individual computer users, system administrators, technology developers, and business executives. These goals would be accomplished by empowering "all Americans to secure their own parts of cyberspace," the fostering of training and education programs, and the promotion of private-sector support for widely recognized professional cybersecurity certifications. Yet again, the emphasis is placed on creating a public and private sector more capable of defending its own cyber assets through voluntary incentives, rather than on direct governmental defensive actions.

The fourth priority of the NSSC is securing governments' own cyber spaces and "leading by example" through various security initiatives in addition to the encouragement of state and local governments to protect their critical infrastructures. It recognizes that while governments administer only a miniscule portion of the nation's critical infrastructure computer systems, still governments at all levels perform essential services in the fields of agriculture, food, water, public health, emergency services, and many others, and that all rely on cyberspace for their delivery. The plan calls for the security of these electronic systems through, first, continuous assessment of threats and vulnerabilities to federal systems, second, authenticating and maintaining authorized users of federal systems, and third, improving security in government outsourcing and procurement. Furthermore, it stipulates that all federal wireless local area (Wi-Fi) networks be made secure, and encourages state and local governments to take these same actions, and to participate in information-sharing and analysis centers with similar governments.

The fifth and final priority mentioned in the NSSC is fostering a system of National Security and International Cyberspace Security Cooperation to promote a "global culture of security." The plan acknowledges that because of the Internet's decentralized and "borderless" architecture, the only way to safeguard and defend America's critical systems and networks is to require "a system of international cooperation to facilitate information sharing, reduce vulnerabilities, and deter malicious actors." The means for accomplishing this international security effort include strengthening counterintelligence efforts, improving response capabilities to attacks, and establishing international watch-and-warning networks to detect attacks as soon as they emerge. Once again, emphasis is placed on working with private industry to facilitate partnerships among international public and private sectors. Interestingly, the plan also calls for encouraging other nations to accede to the Council of Europe Convention on Cybercrime, despite the fact that the United States itself is not a party to it.

Ultimately, the NSSC explicitly recognizes that "the federal government alone cannot sufficiently defend America's cyberspace." It calls for voluntary public-private partnerships and for "every American who can contribute to securing part of cyberspace [being] encouraged to do so." It also acknowledges that "many issues could not be addressed in detail, and others are not yet ripe for national policy."[61]

In the immediate weeks following the release of the draft version of the NSSC, Congress approved the creation of the DHS and assigned to it many agencies already active in cybersecurity. The final version of the NSSC "reflects those changes." Congress also passed, and the president signed, the *Cyber Security Research and Development Act,* which authorized $700,000 in new funding for cybersecurity research and development, primarily to be dispersed in the form of academic fellowships, grants, and the establishment of a program of assistance to institutions of higher education that enter into partnerships with for-profit entities.[62]

Additionally, in June 2003, recognizing the importance of the Protocol layer to cybersecurity, the U.S. Defense Department formally issued a directive mandating a transition to the IPv6 protocol, in accordance with an explicit recommendation in the NSSC, citing the need for more IP addresses being necessary for enhancing national cybersecurity and for military combat operations in the war on terror.[63]

In the end, both the policymaking process and the policy design of the NSSC reflect the importance of our four-layer model. The policymaking process, though instigated by the federal government, nevertheless incorporated many of the governing actors revealed in our political architecture of the Internet—the major telecommunications providers who own the Infrastructure, the largest private commercial software development firms operating at the Applications layer, and many of the largest private ISPs and website operators who hold authority over the Content. Notably absent from the process: the international engineering consortium groups that govern the Protocols (however, the importance of the protocols themselves are explicitly recognized in the NSSC), as well as independent non-affiliated software developers and website operators. Large private commercial firms were clearly dominant in this policymaking process.

Likewise, the policy design of the NSSC also reflects the importance of our four-layer model. It explicitly cites protection of the cyber Infrastructure as its main priority, and also calls for the widespread utilization of more secure technical Protocols and patches for software applications. Its overarching design emphasis on voluntary public-private partnerships is extremely significant and is its most controversial element. While, certainly, it is partially the result of the Bush administration's political bias against direct governmental mandates regulating the private commercial sector, it is also an implicit acknowledgment that, first, the federal government does not actually have the governing authority on its own to create policies that constrain and enable behavior on the Internet with intentional effects; partnerships with private-sector actors who hold governing authority at each layer are not only desirable, but fundamentally required. Second, as a consequence, that this decentralization of governing authority across layers makes voluntary measures, as opposed to direct mandates, not merely a design choice, but really the only option in the context of the Internet's decentralized political architecture that we have constructed.

IMPLEMENTATION AND THE EMERGING BUREAUCRATIC REGIME

- Policy Achievements
- Inadequate Funding
- Critiques of Private-Sector Implementation Efforts

- Conflicting organizational roles within DHS
- High turnover rate among top cybersecurity officials at DHS

In public policy literature, implementation theory is often broken down into the categories of top-down and bottom-up approaches. Among the top-down theorists is the central notion that successful implementation depends upon linkages between different organizations and departments at the local level, and that the degree of cooperation between these agencies will determine the "implementation deficit."[64] A policy is created and its success simply depends on how well local agents carry it out. This is supported by parallel efforts that seek to measure "outcome performance"[65] and to "structure implementation" by recognizing the existence of a feedback process.[66] There is also an acknowledgment that certain political processes are inevitable and therefore implementation must include "scenario writing"—structuring the game the right way to achieve desired outcomes.[67]

On the other hand, bottom-up theorists focus on the complexity between organizations, or networks, in ways that do not necessarily privilege any specific actor or set of actors.[68] Central to this approach is the idea that implementation does not involve the advancement of public service ideals as much as we'd hope, but rather reflects the frustrating processes that lead to practices which enable officials to cope with the pressures they face.[69] In response, the notion of "backward-mapping" attempts to focus on individual actions as a starting point, enabling such actions to be seen as responses to problems or issues in the form of choices between alternatives.[70] Compromises typically arise between people and organizations during the process, thus politicizing the policy-action relationship.[71] It should be noted, however, that some scholars have criticized this bottom-up emphasis on political processes and compromises during implementation as obliterating the distinction between policy formulation and implementation.[72]

What the case study of the NSSC illustrates is a purposive melding of policy formulation and implementation. Policymaking theories such as "multiple streams"[73] or "punctuated equilibrium,"[74] while useful in conceptualizing the problems or needs which sometimes initiate agenda-setting and policy change, prove less reliable as predictive indicators when analyzing a highly technical and nascent field like cybersecurity which lacks a substantial historical record. Instead, incrementalism and the classic notion of "muddling-through"[75] are more accurate descriptions of the policy course that has transpired. In fact, the cybersecurity case study highlights the manner in which implementation is actually a form of incremental policymaking within bureaucratic institutions.

Overall, implementation of the NSSC has been problematic. Since the policy was created, implementation of some of its tenets has led to positive results—notably the emergence of numerous private-sector initiatives and industry alliances—while others have encountered various difficulties—namely, a scarcity of resources, confusion and conflict within the Executive bureaucracy, and a high turnover rate among top administrative officials.

In terms of policy achievements, numerous private-sector initiatives have been established such as the Internet Security Alliance (ISA), which was established as a collaborative effort between Carnegie Mellon's Software Engineering Institute (SEI), its Computer Emergency Readiness Team Coordination Center (CERT/CC), and the Electronic Industries Alliance (EIA)—a federation of trade associations which sought to provide a forum for information-sharing on cybersecurity issues.

Additionally, as stated in Microsoft's Chief Security Strategist Scott Charney's, testimony before the House Subcommittee on Governmental Efficiency, Financial Management, and Intergovernmental Relations, implementation efforts were undertaken seeking to strengthen law enforcement's capabilities in deterring cybercrime through both an expansion of their legal powers as well as increasing their funding for personnel and training, heightening penalties for cybercrime offences, increased funding for cybersecurity research and development (R&D), enhancing cross-jurisdictional cooperation among law enforcement agencies for investigating cyber attacks, and providing more clarity for which governmental agency should take the lead in responding to a specific attack and what legal authorities will guide an investigation.[76]

Furthermore, the U.S. Chamber of Commerce outlined five major private-sector cybersecurity accomplishments. Notable among these were a National Cyber Security Summit, after which five participating organizations—the Business Software Alliance (BSA), the Information Technology Association of America (ITAA), TechNet, the Chamber of Commerce, and DHS—decided to formalize and sustain the National Cyber Security Partnership (NCSP) which has since grown to include over 30 major organizations.[77] It has also provided training to numerous small businesses, as well as published a series of articles on "Common Sense Guides to Cyber Security" for large and small businesses.[78]

Large-scale testing of security vulnerabilities has also offered positive results in terms of local, state, federal, and international governmental cooperation. The DHS's subdivision—NCSD—has sponsored and carried out four large-scale national cybersecurity exercises. Named Cyber Storms I-IV, spanning from 2006–2012, these exercises simulated a large-scale national cyber attack in order to assess coordinated federal responses. Participants included federal, state, local, and international governments, including Australia, Canada, New Zealand, and the United Kingdom, in addition to private-sector actors from the Information Technology, Transportation (Rail and Pipe), and Chemical sectors, along with multiple Information Sharing and Analysis Centers (ISACs).[79]

Since the passage of the NSSC, the DHS has also begun funding numerous R&D programs, albeit, not to the extent that some proponents argue is required. Protected Repository for Defense of Infrastructure Against Cyber Threats (PREDICT) is a program begun in 2004 aimed at getting large private-sector infrastructure companies to volunteer real-world incident data that researchers can use to test prototype security products. The agency also spearheaded a new vender-neutral cybersecurity test bed, known as DETER, for Cyber Defense Technology Experimental Research, in order to help develop next-generation security technologies for the nation's critical infrastructure. DHS also formed an ad hoc government/industry steering committee to study and develop security pilot projects for the Internet's DNS System. The goal is to develop pilot projects to study specific threats and vulnerabilities to the DNS System, including loss of service due to a denial-of-service attack, hijacking, and a loss of coherence due to the existence of unauthorized root servers and top-level domains. Additionally, with another nod to the importance of the Protocol layer, its Border Gateway Protocol steering committee prepared R&D pilot projects to develop secure protocols for the routing infrastructure that connects ISPs and subscriber networks.[80]

However, the actual funding of the NSSC's proposals has been piecemeal, and this has greatly hindered its implementation. Despite calling for "federally funded near-term IT security research and development," grants to universities for the training of professionals, and other measures that add up to significant costs, Congress has largely funded associated proposals as parts of larger DHS (and other) bills, rather than as one comprehensive funding package dedicated specifically to cybersecurity. As a result, federal funding for cybersecurity has been notably porous in the years since the NSSC's creation.

Funding Figures

As a result of inadequate funding, implementation of the cybersecurity measures called for in the NSSC has been demonstrably impacted. Congress, after passing the Federal Information Security Management Act (FISMA) in order to supplement the NSSC, requires all federal civilian agencies to test their systems for cyber-vulnerabilities and report annually to Congress on their progress.[81] The resulting annual FISMA report cards have revealed the common frequency by which many government agencies receive extremely poor marks on the official report card—for example, generating an average grade of 67.3 percent for 2004, an improvement of only 2.3 percentage points over 2003.[82]

Table 7.3 Funding Figures for DHS and NCSD: 2002–2008

Fiscal Year	*Federal Budget*	*DHS Budget (Department of Homeland Security)**	*NCSD Budget (National Cyber Security Division)*
2002	$2.0 trillion	$19.5 billion	—
2003	$2.2 trillion	$34.2 billion	—
2004	$2.3 trillion	$36.2 billion	—
2005	$2.4 trillion	$40.2 billion	—
2006	$2.7 trillion	$41.1 billion	$73 million**
2007	$2.77 trillion	$42.7 billion	$93 million***
2008	$2.9 trillion	$46.4 billion	$115 million****

* All DHS Budget figures come from the U.S. Department of Homeland Security website – DHS Budget. Retrieved on June 6, 2013 from http://www.dhs.gov/dhs-budget.
** "DHS Budget FY06," U.S. Government Printing Office website. Retrieved on July 31, 2008 from http://www.gpoaccess.gov/usbudget/fy06/pdf/budget/dhs.pdf.
*** "DHS Budget FY07," U.S. Government Printing Office website. Retrieved on July 31, 2008 from http://www.gpoaccess.gov/usbudget/fy07/pdf/budget/dhs.pdf.
**** Retrieved on July 31, 2008 from http://www.gen.com/online/vol1_no1/46080-1.html.

Such funding inadequacies have not gone unnoticed. In 2007, after a six-month stretch in which two highly prominent events occurred—a major cyberattack on Estonian government web sites and intrusions into the Defense Department's unclassified network—President Bush requested that Congress "immediately move $152 million into cybersecurity programs for fiscal 2008." Out of this money, $115 million would go toward enhancing DHS' ability to implement the Einstein program, which monitors network gateways for traffic patterns that indicate the presence of computer worms or other unwanted traffic, administered by US-CERT. Jeff Carter, a spokesman at DHS' Directorate of National Protection and Programs (DNPP), also said that the funding "will also increase our investigative capabilities [and] reduce multiple access points and points of vulnerability while ensuring government cyber centers stay connected and focused on detecting potential attacks." The Justice Department, meanwhile, would receive $39 million "to help the FBI investigate incursions into federal networks, increase intelligence analysis and provide technical tools for investigations."[83]

Some have argued that the cause of implementation's difficulties lies in the policy design itself or even in the exclusion of certain types of actors from the policy process. Michael Rasmussen (the aforementioned V.P. for the Information Systems Security Association) asserted in an interview in December 2003 that what had stalled progress in implementing national cybersecurity policy was:

> the direction the Federal Government is heading with the Public-Private sector cooperation on information security. We are only now seeing a direction in this area. A direction was being clearly laid by Howard Schmidt and Richard Clarke,

but the creation of DHS and integration of this component into it stalled it big time.[84]

Some of the critiques that were made during the public commentary phase of the policy process have subsequently been echoed in the years since the NSSC's passage. In December 2003, Robert Vamosi of About.com wrote,

> At the National Cyber Security Summit, held last Wednesday in Santa Clara, Calif., government officials praised the progress they've made thus far. However, it's interesting to note that the 300 invited guests at the closed-door sessions did not include many noted individuals within the security community, giving the summit a distinctly pro-business skew.[85]

Private-sector implementation efforts have often struggled as well. Leading technology companies like Microsoft, for example, have established their own enhanced security programs such as the Trustworthy Computing Initiative in order to certify the security of its products as well as its partners in the Initiative. However, many such private-sector efforts are often scrutinized for their potential ulterior motives. For instance, Microsoft's insistence that Digital Rights Management (DRM) software is essential to the Trustworthy Computing Initiative has led many detractors to question the entire project, since many view DRM as an attempt by Microsoft to not only protect, but also control, media content on users' computers. Furthermore, the open-source Community has also expressed concern that a trustworthy computing implementation will require authenticating programs as well as content. Such a system could potentially be used to hinder the progress of non-Microsoft software and operating systems—leading to allegations of anticompetitive behavior.

One of the greatest obstacles to the NSSC's implementation in its early years was the high turnover rate among top officials and organizational conflict within the Executive bureaucracy. A brief history of the "Cybersecurity Czar" position illustrates this point.

On October 12, 2001, only one month after the 9/11 attacks, President Bush named Richard Clarke to the new position of Special Advisor to the President for Cyberspace Security, immediately making him the president's top advisor on cybersecurity issues. Clarke had previously been a member of the Clinton and George H.W. Bush administrations and a NSC staffer. He was also the president's counterterrorism coordinator at the time of the 9/11 attacks. In this new position, Clarke reported directly to Homeland Security Office Director Tom Ridge and National Security Advisor Condoleezza Rice.

Six days later, President Bush issued an Executive Order creating the CIPB headed by Clarke, comprising 26 federal agencies. The Board "was formed to protect the information infrastructure controlling everything from financial systems to the power grid to telephone and internet communications." By early November, Clarke made a 10-day trip to Silicon Valley to solicit suggestions on cybersecurity from 18 top executives, including Cisco CEO John Chambers and Symantec CEO John Thompson. The Board, working closely with industry executives, was responsible for drafting the NSSC.

Clarke resigned from his post in January 2003 following damage wreaked by an Internet worm that struck hundreds of thousands of computers worldwide, slowing e-mail systems and other cyberspatial activities. The "SQL Slammer Worm" (also known as Sapphire and Helkern) also crippled 911 emergency centers, prevented many customers of Bank of America Corp. from withdrawing money from ATM machines, and Countrywide Financial Corp., Microsoft and American Express Co. also reported problems. Clarke stated that the Bush administration was "not taking cybersecurity seriously." He also famously told security experts at the RSA Data Security Conference in 2002, after citing statistics that indicated less than 0.0025 percent of corporate revenue on average was being spent on information technology security, that "If you spend more on coffee than on IT security, then you will be hacked. What's more, you deserve to be hacked."[86]

Clarke was followed by Howard Schmidt, the Vice-Chair of the CIPB and a former chief of security at Microsoft, who then resigned after only three months in the position. In an e-mail sent to his staff and other industry officials immediately following his resignation, Schmidt stated that many of his responsibilities had been shifted to the DHS, and following Richard Clarke's precedent, also warned of future cyberattacks in calling for officials to ensure that cybersecurity not be reduced to a "second-tier issue."[87]

Succeeding Schmidt was then Rand Beers, who quit after only one month on the job in order to join the presidential campaign of Senator John Kerry. Beers, who served on the NSC under four presidents, charged the Bush administration with underfunding and taking little action to improve cybersecurity. This, he claimed, led to a situation of "policy constipation" where "nothing gets done."[88]

Following Beers was Amit Yoran, a former software executive from Symantec Corp., who then, in October 2004 "informed the White House about his plans to quit as director of the NCSD and made his resignation effective at the end of Thursday, effectively giving a single's day notice of his intentions to leave." After one year in office, Yoran cited frustration with cybersecurity's "low priority" at DHS, and "privately described [those] frustrations in recent months to colleagues in the technology industry, according to lobbyists who recounted these conversations on condition they not be identified because the talks were personal. As cybersecurity chief, Yoran and his division had an $80 million budget and 60 employees to carry out the dozens of recommendations in the NSSC."[89]

With Yoran's departure, that made for four resignations within a year and a half.

In February 2005, the President's own Information Technology and Advisory Committee (PITAC) issued a highly critical report titled, "Cyber Security: A Crisis of Prioritization," in which it described the short-term, and severely flawed, strategies that had thus far been pursued in cybersecurity in accordance with the NSSC. The report urged several changes be made to the policy such as significant increases in funding for research, recruitment and retention of cybersecurity researchers and professionals, and providing a "rapid transfer" of federally developed cutting-edge cybersecurity technologies to the private sector.[90]

As a result, in July 2005, the more powerful post of "cybersecurity czar" was officially created as a part of a broad reorganization at the DHS. In his "six-point agenda," DHS Secretary Michael Chertoff elevated the position of "cybersecurity chief" several levels up the agency's organizational chart by creating this new position officially titled, "Assistant Secretary for Cyber and Telecommunications Security." This new assistant secretary would report directly to the Undersecretary for National Protection and Programs Directorate, one of three top-level officials who answer directly to Chertoff. Several tech-oriented trade groups, including the ITAA and the Cyber Security Industry Alliance (CSIA), founded in 2004 by security firms such as Symantec, McAfee, RSA Security, Check Point, and Internet Security Systems, had repeatedly called for the creation of an assistant secretary position "in order to raise the profile of cybersecurity issues at DHS."

However, the new position of cybersecurity czar would remain vacant for over a year. This prolonged vacancy drew criticism from politicians and technology industry groups alike.

The CSIA's Executive Director Paul Kurtz made regular appearances before House and Senate subcommittees to submit testimony stressing the need for a private-sector approach to improving cybersecurity without major government intervention, and was integral in calling for the appointment of an Assistant Secretary for Cybersecurity and for Congressional enactment of data security legislation in 2006.[91]

The BSA, whose members include Apple Computer, Cisco Systems, Dell and Microsoft, submitted a three-paragraph letter to Chertoff in July 2006 pressing for an appointment to the position "in the near future."[92]

Shortly thereafter in the summer of 2006, DHS finally appointed a new cybersecurity czar. Gregory Garcia, formerly a vice president of the aforementioned ITAA, was named to the post with a background in both computer security and business.

Then, in a move that surprised many observers, in March 2008, President Bush signed a new directive to expand the intelligence community's role in monitoring Internet traffic following a surge in the number of attacks on federal agencies' computer systems. The directive, whose content details were classified, authorized the NSA to monitor the computer networks of all federal agencies. This directive was part of a new strategic initiative calling for a task force headed by the Office of the Director of National Intelligence (ODNI) to coordinate efforts to identify the sources of cyberattacks, while DHS would work to protect the computer systems, and the Pentagon would devise strategies for counterattacks.

In this strategic initiative known as the Comprehensive National Cybersecurity Initiative (CNCI), the creation of a new multiagency, multiyear plan was set forth, engendered by "Homeland Security Presidential Directive 23"—whose details are also classified—which lays out 12 steps to securing the federal government's cyber networks, and establishes the NCSC.[93]

The CNCI initiative immediately raised a number of questions regarding the cybersecurity czar. To head the new interagency, the president appointed Rod A. Beckstrom, a Silicon Valley entrepreneur who started Twiki.net, a company which provides collaboration software for businesses, and author of *The Spider and the Starfish: The Unstoppable Power of Leaderless Organizations*, as his new top-level advisor based within DHS, reporting directly to Secretary Chertoff. Beckstrom's appointment raised new questions as to what was then the role of the official cybersecurity czar, Gregory Garcia.

The CNCI had been cloaked in secrecy, as DHS initially withheld all information about the new agency, the NCSC. It was signed by President Bush in January 2008, and "there are rumors that Congress will be asked to come up with as much as $30 billion over coming years." Senators and Congressional Representatives were only informed behind closed doors after DHS officials were called to testify or make budget requests, and DHS officials claimed the entire program was classified. There were also additional reports that the NSA, Central Intelligence Agency (CIA), FBI would cooperate on monitoring and share information via the new NCSC, the aim of which was also classified.[94]

This secrecy ultimately prompted Senators Joe Lieberman (I-CT) and Susan Collins (R-ME), the Chair and ranking Republican of the Senate's Homeland Security and Governmental Affairs Committee, to send a letter to DHS Secretary Michael Chertoff raising no less than 17 questions about the NCSC and its activities. The Senators' questions centered on issues ranging from the project's secrecy to its heavy reliance on contractors to the lack of involvement by the private sector.[95]

As it turned out, Rod Beckstrom did not last long as director of the NCSC either. In March 2009, he resigned his position in part as a show of resistance to the NSA's interest in taking a dominant role in cybersecurity. Beckstrom said in an interview that he believed an intelligence service that is supposed to focus on foreign targets should not be given so much control over the flow of information within the United States government. He was quoted as having "very serious concerns about the concentration of too much power in one agency," and that he feared that the NSA's push for a greater role could "give it the power to collect and analyze every e-mail message, text message and Google search conducted by every employee in every federal agency."[96]

Beckstrom's resignation illustrated, first, just how divisive the cybersecurity issue had become among federal agencies, as demonstrated by the conflicting roles between the NCSD and the NCSC, and, second, that the high turnover rate among top bureaucratic officials remained ongoing.

CYBERSECURITY POLICY AND POLITICS IN THE OBAMA ADMINISTRATION

With the ushering in to power of the Obama administration, the policy course set forth by the Bush administration—characterized by public-private partnerships and strictly voluntary measures being promoted to enhance cybersecurity—was largely kept intact. This was a case of path dependency having already taken hold.

It wasn't until April 2012 that a notable change in policy direction was even expressed. President Obama's chief counterterrorism advisor, John Brennan, publicly made the case that the strictly voluntary approach had become "a risk that the American people cannot afford to take."[97]

His public opposition to the entrenched voluntary approach echoed the principles set forth in the Senate's newly released *Cybersecurity Act of 2012*. This plan called for the federal government to set minimum cybersecurity performance standards—after garnering industry input—and companies who worked on or operated the nation's critical cyber assets would be required to meet them. Those companies who fell short would be "directed" to tighten up their cybersecurity practices. Exactly how they would do so—for example, behind a firewall or a stand-alone network—would be up to the company.[98]

Despite this bill being introduced in the Senate, neither it nor any other significant piece of cybersecurity legislation was actually passed by Congress during the first term of the Obama administration. As a result, the status quo remains in place—meaning that the Bush administration's policy course based on public-private partnerships and strictly voluntary measures being adopted continues to define U.S. national cybersecurity policy to date.

However, under the Obama administration, the politics of cybersecurity have become more crystallized. Partially in response to years of frustration stemming from organizational conflict within DHS, several other governmental actors have become highly prominent in framing the issue and setting the agenda.

As already stated, the leading Senate committee dealing with cybersecurity policy is the *Senate Committee on Homeland Security and Governmental Affairs*. Chaired by Sen. Joe Lieberman with Republican ranking member Sen. Susan Collins, this committee has a history of introducing bills like 2010's the *Protecting Cyber Space as a National Asset Act* (S. (3480), which passed out of committee but was never debated on the Senate floor.[99]

It was this committee that was also responsible for producing the *Cybersecurity Act of 2012* (S. (2105) which was introduced by Senators Lieberman, Collins, Jay Rockefeller, and Diane Feinstein. The Act was the result of months of negotiations with other committees of jurisdiction—namely, the energy, financial services, and chemical industries; national security and privacy and civil liberties groups; and a number of other government agencies.[100]

Meanwhile, by its own admission, in the House of Representatives, at least nine committees have some significant jurisdictional claim on cyber issues. These include Appropriations, Oversight and Government Reform, Armed Services, Judiciary, Financial Services, Homeland Security, Science, Space, and Technology, Energy and Commerce, and the Permanent Select Committee on Intelligence.[101]

In order to determine, not only which among these House committees are most important to cybersecurity policy, but also the Executive Branch turf wars involved, it is instructive to follow the action in the House's counterpart to the *Cybersecurity Act of 2012*.

Two competing bills were introduced in the House. The Rogers-Ruppersberger bill (H.R. (3523)[102] was the product of the *Select Committee on Intelligence* and was passed out of that committee by a 17–1 vote. It would give a leading role to the Director of National Intelligence, making him responsible for establishing procedures to broadly share cyber threat information with the private sector.

The other, H.R. 3674, is known as the *PRECISE Act* (Promoting and Enhancing Cybersecurity and Information Sharing Effectiveness Act).[103] It was authored by Congressman Dan Lungren (R-CA), has bipartisan support in the *Homeland Security Committee*, and would give a leading role to the Secretary for Homeland Security. DHS would be responsible for maintaining a clearinghouse of cyber threat information and disseminating that information broadly within the federal government and to the private sector.

This effort at revamping U.S. national cybersecurity policy in 2012 clearly illustrated that the principal Congressional actors are, in the Senate, the Homeland Security and Governmental Affairs committee, and in the House, the Select Committee on Intelligence as well as the Homeland Security Committee. Furthermore, the two competing bills in the House highlight the ongoing tension between DHS and DoD/NSA as to who will be the primary intermediary between the federal government and the private sector.

Certainly, there remain other subsidiary yet highly important cybersecurity actors as well. The military plays an extremely active role in formulating cybersecurity strategies, engaging in research and development of cyber technologies, and running scheduled exercises that simulate attacks and test vulnerabilities.

Foremost, *U.S. Cyber Command (CYBERCOM)* was established in 2010 and is headed by General Keith Alexander, who is also the NSA director, and reports to the U.S. Strategic Command. Service elements of CYBERCOM include the *Army Cyber Command, the 24th Air Force, the Navy's 10th Fleet Cyber Command, and the Marine Forces Cyber Command.*[104]

In part as a consequence of its relationship with CYBERCOM (although not limited to it), *the NSA* also plays a highly prominent role. In addition to various other duties, the NSA chief is in command of CYBERCOM, as well as those aforementioned subsidiary single-service cyberwar units such as the 24th Air Force, Navy 10th Fleet, etc.[105]

One example of NSA activity on the cybersecurity front was when news emerged in 2010 that the agency had set established a secret program called "Perfect Citizen" that was intended to set up monitoring equipment on networks deemed to be of national security importance, perhaps including those of utility companies. In theory, this would allow the NSA to know when attacks were happening, rather than relying on private companies to report it. However, the NSA denied that there was any monitoring involved.[106] Nevertheless, it is clear that the NSA plays a key role both in developing and deploying cyberdefense mechanisms.

Finally, as we have discussed, *DHS* has been central to U.S. cybersecurity policy since its inception, and remains so. It has been charged with being the primary coordinator of information related to cyberattacks, and is still considered the first liaison between the public and private sectors. As we have explored in great detail, conflicts between different DHS departments—the NCSD and the NCSC—coupled with a high turnover rate among top officials have defined the Department's early years, at least in the context of cybersecurity.

In March 2013, the Obama administration began publicly staking out a more proactive (or aggressive) cybersecurity strategy. The president issued a new Executive Order on the matter, although it merely encouraged greater information-sharing with the private sector—an extension of the existing policy. Also, Administration officials communicated to China's new president, Xi Jinping, that "the volume and sophistication of Chinese cyberattacks ha[d] become so intense that they threaten[ed] the relationship between Washington and Beijing." Their solution, though, was more timid, suggesting that Chinese diplomats help establish "acceptable norms of behavior in cyberspace." However, in a marked contrast to previous policy, General Keith Alexander, head of both the NSA and CYBERCOM, testified before the House Armed Services Committee about the Administration's intention to establish 13 teams that could launch *offensive* cyberattacks in retaliation if the U.S. were ever hit with a major attack.

In the meantime, the aforementioned *Cybersecurity Act of 2012*, which proposed replacing the strictly voluntary measures in the NSSC with government mandates for companies who worked on or operated the nation's critical cyber assets, was defeated in the Senate by a vote of 52–46.

The extent to which these new strategies of offensive cyberwarfare and governmental mandates will actually become adopted, as of this writing, remains uncertain.

CYBERSECURITY POLICY IN ACTION: WHAT ACTUALLY HAPPENS IN THE FACE OF A CYBERATTACK?

To see how the *National Strategy* policy, its implementation, and the political architecture surrounding the entire issue all come together in terms of our four-layer model, it is instructive to trace the process of what happens when there is an actual computer virus outbreak.

As set forth in the NSSC, U.S. cybersecurity policy is based on two main principles which parallel broader national security objectives: (1) prevention and (2) response.

As a direct result of the policy design of the NSSC relying foremost on public-private partnerships, both prevention and response strategies rely first on the private sector to voluntarily implement measures to protect their own

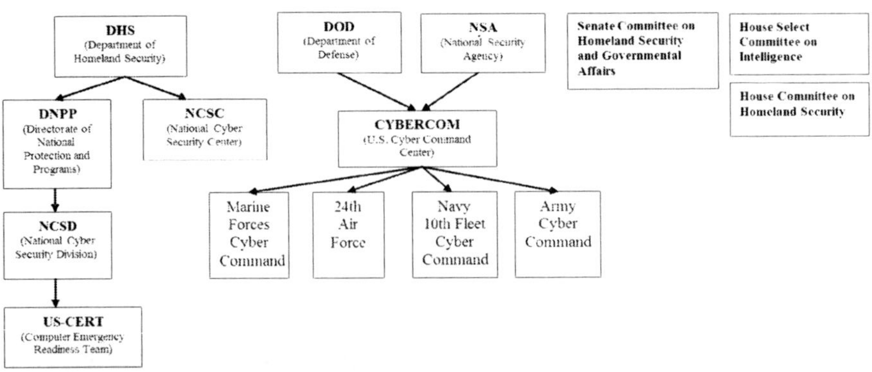

Figure 7.1 The Federal Cybersecurity Regime

cyber assets, and only after such private measures have run their course does the federal government take direct action.

Computer viruses have existed in some form since the early 1970s when the "Creeper" virus was first detected on the ARPANET, and they have been a recognized threat to information systems since Len Adleman first coined the term "virus" in connection with self-replicating computer programs in 1983.[107]

A notable turning point as to how viruses relate directly to national cybersecurity policy came in 1999 when the "Melissa" virus wreaked havoc on computer systems around the world. Melissa exploited the macro programming language used by Microsoft software applications to disable certain features within Microsoft Word, then sent copies of the infected document to up to 50 other addresses using compatible versions of Microsoft's Outlook e-mail program. Finally, the virus modified the Word software so that it would subsequently infect any document that the user might open and close. If these documents were shared, the virus would be spread further.

The U.S. General Accounting Office (GAO) shortly thereafter published the testimony of Technical Director Keith Rhodes before the Congressional Subcommittee on Technology, Committee on Science. In his testimony, Rhodes highlighted that the Melissa virus was important because it demonstrated (1) how quickly viruses can proliferate "due to the intricate and extensive connectivity of today's networks," and how difficult it was to launch effective countermeasures, (2) how hard it is to trace a virus back to its source, (3) how vulnerabilities in commercial-off-the-shelf (COTS) software products could be exploited to affect vital federal control systems which increasingly had grown dependent on such software, (4) the lack of an effective agency or government-wide processes for reporting and analyzing the effects of computer attacks, and (5) how individual computer users do a good job of protecting their systems when they are made aware of computing risks and attacks.[108]

Furthermore, Rhodes outlined the federal government's role in mitigating the effects of computer virus outbreaks as assuming leadership in coordinating information-sharing with the private sector:

> It is imperative, therefore, that federal agencies and the government as whole swiftly implement long-term solutions to protect systems and sensitive data. It is also critical that the federal government establish reporting mechanisms that facilitate analyses of viruses and other forms of computer attacks and their impact. Our Information Security Best Practices guide offers a good framework for agencies to follow, but sustained government-wide leadership is needed to ensure that executives understand their risks, monitor agency performance, and resolve issues affecting multiple agencies.[109]

Institutionally, when a virus hits, the primary federal agency responsible for mitigating its effects is US-CERT. US-CERT is the operational arm of the NCSD at the DHS. It is a public-private partnership charged with providing "response support and defense against cyber attacks for the Federal Civil Executive Branch and information sharing and collaboration with state and local government, industry and international partners." One of its main roles is also to "disseminate reasoned and actionable cyber security information to the public."[110]

Worldwide, there are more than 250 organizations dealing with cybersecurity response that use the name "CERT." Although there is some level of coordination with these groups, US-CERT is independent of them. When DHS created US-CERT, it called upon the CERT/CC established at Carnegie Mellon University to contribute their expertise, and it is through US-CERT that DHS and the CERT/CC continue to work jointly on cybersecurity activities.[111]

In order to enhance response systems, the Protected Critical Infrastructure Information (PCII) Program was established as a result of the *Critical Infrastructure Information Act* of 2002.[112] Its purpose is to encourage members of the private sector to voluntarily submit security information about vulnerabilities to DHS by enabling them to submit such information confidentially with the assurance that it will be protected from public disclosure.[113]

The other important institutions, aside from US-CERT, include the Multi-State Information-Sharing and Analysis Center (MS-ISAC) which focuses on states and local governments,[114] CERT/CC at Carnegie Mellon University, the National Institute of Standards and Technology (NIST)—Computer Security Division which focuses on standards and technology-based approaches,[115] and the Forum of International Response Security Teams which is global in scope.[116] There is also within the U.S. House of Representatives the Subcommittee on Emerging Threats, Cybersecurity, and Science and Technology.[117] It is a part of the Committee on Homeland Security.

In substantive terms, when it comes to computer virus outbreaks, organizations are expected to take certain steps *before* an outbreak occurs. Clearly, this path pursues the objective of prevention.

According to the NIST Incident Handling Guide,[118] organizations, both public and private, need to create formal incident response capabilities. These can vary greatly in terms of both type and scale, depending on the characteristics of who is implementing them.

Federal agencies are required by law to report incidents to the Federal Computer Incident Response Center (FedCIRC) office within DHS's Information Analysis and Infrastructure Protection Directorate (IAIP).

The private sector is encouraged to take several preventative steps as well, although these measures are voluntary and considered best practices. NIST recommends that institutions create formal policies and procedures to be well-prepared to handle incidents when they occur. Examples include internal policies that explicitly state when to share information with outside parties, creating team models and selecting the best personnel, and listing dependencies within organizations.

Additionally, organizations need to take steps to effectively secure their networks, systems, and applications. From a procedural perspective, this means implementing technical measures for detection and analysis, and preselecting strategies for containment, eradication, and recovery—typically accomplished through software applications and network management.

Organizations are further instructed to document their guidelines for interactions with other organizations, emphasize the importance of incident detection and analysis to their employees, and create written guidelines for prioritizing incidents. These priorities ought to be determined by the (1) criticality of the affected resources (e.g., public web servers, user workstations), and (2) the current and potential technical effect of the incident (e.g., root compromise, data destruction).[119]

It is important to stress that the overwhelming majority of such preventative techniques are expected to occur in the private sector, including among individual home users, and are completely voluntary in nature. While government agencies are required to take such steps, U.S. cybersecurity policy, when it comes to the prevention of widespread security incidents, essentially is a statement of encouragement and advice for private network operators and individual PC-users to protect themselves, and it is this upon which the stability of the system depends.

It follows, then, that organizations are also expected to take specific steps *after* an outbreak occurs. This path pursues the complementary objective of response.

The steps that organizations should take in response vary depending on the type of security breach that has occurred. While the potential courses of action are immensely numerous, the intention is that the decision-making processes for containing a security incident will be far easier if recommended actions are predetermined.

For Denial-of-Service attacks, contact should be made with one's ISP as well as their second-tier service providers. Because the issue with DOS attacks is an overload of network-based traffic, ISPs are critical for filtering or limiting that traffic, and they have several means for doing so such as blocking source IP addresses or setting a maximum limit for incoming traffic. Relocating the target of the attack by moving it to another IP address is another viable option if a particular host is being targeted.

For malicious code incidents, the priority is to prevent the malicious code from spreading any further. If the infected device or system is not critical, it should be disconnected from the rest of the network immediately. Toward the same end, all preventative measures should be re-executed on devices that may not yet be infected, like running antivirus software, configuring e-mail servers to block suspicious files, limiting the use of software applications with file transfer capabilities such as IRC and instant messenger clients, etc. In response, contact should then be made with antivirus software vendors to alert them of the unknown malicious code that their software isn't able to identify.

For unauthorized access incidents, the key is to develop a strong layered defense with multiple security layers existing between unauthorized users and critical resources. Since unauthorized access typically depends on initial reconnaissance missions, organizations are recommended to use network-based monitoring software (such as file integrity checkers) and centralized log servers to detect intrusions before they gain administrator-level access. If that occurs, however, the response procedure ought to isolate and then disable the affected system, eliminate the attacker's route into the environment (whether by changing passwords or altering database privileges, most commonly), and finally disabling any user accounts that may have been compromised in the attack. Since attackers often install rootkits, handlers ought to reinstall the operating systems themselves from scratch. Administrators need to also be aware of various laws that require disclosure of security breaches, particularly pertaining to personally identifiable information.

For inappropriate usage incidents, preventative measures to be taken include deploying content-filtering software and logging certain user activities such as FTP commands, Web requests, and e-mail headers. While legal mechanisms (such as nondisclosure agreements, cease-and-desist orders, and laws protecting trade secrets) are central to the response of such an incident, because many inappropriate usage security threats come from within an organization, technical measures can be most effective in governing one's own network. Examples include configuring firewalls, URL filters, e-mail servers, and outbound connections that use encryption protocols.

The common denominator in all of these cases is that, when a security incident occurs, an organization or individual user ought to, first, follow predetermined internal procedures to try to contain and eradicate the problem (usually by deploying software and specialized network tools), second, notify their ISPs and second-tier service providers of the incident, and, third, contact US-CERT if the incident meets predetermined criteria of criticality.

When analyzed through the lens of our four-layer model, what this process illustrates about cybersecurity policy is that the private commercial sector is the "frontline" of national cyberdefense. Protecting the infrastructure from the threat of cyberterrorism is paramount, and it is the private owners of those critical infrastructural assets who have the authority to adopt certain cybersecurity measures, or not. As we have discussed, the federal government does have established regulatory authority at the Infrastructure layer to mandate those firms to meet minimal cybersecurity requirements, but has thus far failed to act upon it.

Also, by tracing the responses of organizations to a cyberattack, it is clear that cybersecurity deployment mechanisms are fundamentally reliant on software applications and technical protocols in both prevention and response, particularly network-monitoring tools and specific antivirus products. Here, too, at the Protocol and Application layers, the private commercial sector is paramount, not only in developing the protocols and software applications that protect digital assets, but also in utilizing such tools to detect cyberattacks, mitigating their effects once discovered, and notifying others of the threat. At these layers, the tools and methods of both prevention and response demonstrate the governing authority of private actors in decision-making.

Finally, this process of what actually occurs in response to a major cyberattack illustrates that the federal government's role in cybersecurity policy is relegated primarily to being a coordinator among numerous private actors who hold governing authority in their own right, and the specifics of which serve to confirm the validity of our political architecture. The government's coordination efforts focus primarily on the large telecommunications firms identified at the Infrastructure layer, the major private commercial software developers at the Applications layer, and the private ISPs and largest website operators at the Content layer. US-CERT is vital to raising awareness about cyberattacks and for information-sharing, but ultimately, U.S. national cybersecurity policy thus far limits the federal government from taking more forceful measures beyond that point. The four-layer conceptual model again proves helpful in contextualizing both the problem stream and solution stream surrounding the issue by framing it in these terms.

WHAT THIS CASE STUDY ON U.S. NATIONAL CYBERSECURITY POLICY DEMONSTRATES: THE PRIMACY OF PRIVATE COMMERCIAL FIRMS

So where does all of this leave us? How can we summarize what exactly is the existing scenario that characterizes U.S. national cybersecurity policy and what are its implications for our four-layer model and its resulting political architecture?

Our examination of the NSSC's policymaking process, its policy design, and its implementation all revealed one commonality—a fundamental dependency on the private commercial sector.

What Actually Happens in the Face of a Cyberattack...

1. Organizations follow predetermined *internal* procedures to try to contain and eradicate the problem.
2. Notify ISPs and second-tier service providers of the incident.
3. Alert anti-virus (and other) software vendors.
4. Contact US-CERT once the incident meets predetermined criteria of criticality.
5. US-CERT will then take the lead in coordinating responses with other important institutions:
    a. MS-ISAC: focuses on state and local governments
    b. CERT/CC: focuses on the more than 250 CERT-certified security organizations
    c. The Forum of International Response Security Teams: focuses internationally
    d. NIST – Computer Security Division: focuses on standards and technology-based approaches
6. The private sector is encouraged to voluntarily submit security information confidentially through the Protected Critical Infrastructure Information (PCII) Program. Federal agencies are required to by law. (This is ongoing and not only during times of crisis.)

Figure 7.2 What Actually Happens in the Face of a Cyberattack

The policymaking process behind the NSSC was heavily influenced at every stage by large private corporations—namely, lobbyists representing IT vendors and various communications, software, and security companies. The policy design that emerged relies almost exclusively on public-private partnerships and strictly voluntary measures being adopted by the private sector in securing their own private cyber spaces.

The implementation process has been bureaucratically characterized by a high turnover rate among top officials and a still-emerging federal regime constituted of numerous agencies with conflicting or overlapping responsibilities. That being said, the *National Cyber Security Division (NCSD)* within the DHS remains the most central government agency in U.S. cybersecurity policy—both in evolving policymaking as well as implementation—charged with coordinating prevention and responses to security challenges among a myriad of private-sector actors.

The NCSD has attempted to implement the principles of the NSSC policy document through four primary means: (1) its *US-CERT* subdivision, which analyzes threats, disseminates warnings, and coordinates incident response activities, and is the single most critical organization for coordinating responses to security threats like widespread computer virus outbreaks; (2) its Cybersecurity Preparedness and the National Cyber Alert System, which assists users to stay prepared by posting current alerts and information; (3) its National Cyber Response Coordination Group, which is made up of 13 federal agencies and, in the event of a nationally significant cyber-related incident, will take the lead in coordinating the federal response, including US-CERT, law enforcement and the intelligence community; and (4) its Cyber Cop Portal, which assists law enforcement capture and convict those responsible for cyber attacks through information-sharing and collaboration with over 5,300 investigators worldwide.[120]

In terms of street-level implementation—what actually happens in the face of a cyberattack—the private commercial sector is clearly paramount. The federal government has some authority in regulating the large telecom firms at the Infrastructure layer, however, because responses to cyberattacks rely so heavily on antivirus (and other security-related) software applications and network management tools, private commercial firms prove to be most vital at the critical Applications layer. Indeed, in their governing authority over the creation and deployment of security software, these firms are the unquestionable "frontline" of national cyber defense.

As we have argued, cybersecurity policy's heavy dependence on the private sector is the direct consequence of the diffusion of governing authority that exists on the Internet between numerous private actors each operating at different layers. It is not so much a policy preference or design choice, but a matter of necessity; an acknowledgment of the decentralized reality, in terms of governance, that has been set forth in our political architecture stemming from the four-layer model. As a result, we have refuted the hypothesis that cybersecurity's failures are mainly attributable to a flawed policy design, since such a design focusing on private-sector arrangements is fundamentally necessary, and instead have pointed to the bureaucratic turmoil occurring in the policy's implementation process for explaining U.S. cybersecurity's failures.

How does the four-layer model assist us in understanding U.S. cybersecurity policy?

Its application to the issue first helps frame the highly complex problem stream which cybersecurity policy is designed to address. Already residing on the national political agenda, the problem definition, as we have illustrated, is comprised of the threats of cyberterrorism, cracktivism, cyberwarfare, and hacktivism to our national economy and critical infrastructure. These problems occur primarily at the Infrastructure and Content layers with respect to cyberterrorism, and the Applications layer with respect to cracktivism and cyberwarfare.

In terms of the policy itself, when viewed through the prism of our four-layer model, the case of cybersecurity highlights that (1) national governments are most relevant at the Infrastructure layer, protecting the network's vital physical hardware, (2) engineering consortia groups like the IETF and IEEE have foremost authority at the Protocol layer, designing and implementing better security within the Internet's technology itself, (3) private network administrators and software developers, in conjunction with the federal government acting as coordinator, are fundamental to the Applications layer, containing the threat of security vulnerabilities becoming more widespread, and (4) private website operators, along with ISPs, are vital to the Content layer, using their governing authority over their own private cyber spaces to monitor, restrict access, and outright remove Web content, not to mention contributing to the discovery of those users who post such content. This confirms the validity of the political architecture of the Internet which we have constructed.

Deconstructing this highly complex issue and analyzing it through the lens of our four-layer model helps to conceptualize its constituent parts in a more meaningful way. Rather than simply asking, "How can the U.S. enhance national cybersecurity?," the question can be broken down into more narrowly targeted questions like "How can the military help prevent the severing of intercontinental undersea cables?"; "What direction should the big three international engineering consortia groups—the IETF, IEEE, and W3C—take on designing the next round of security protocols?"; "What concrete steps can private software developers and network administrators take to mitigate the effects of virus outbreaks within their own systems, and how can a better system of notification be implemented?"; or "What criminal or civil penalties should exist for ISPs or private website operators who knowingly publish information constituting a national security threat?."

The case of cybersecurity is instructive because, as with many other Internet policy issues previously discussed, what are often vague and over-generalized policy dilemmas can be transformed, by using the four-layer model, into more manageable questions with more clearly defined outcome targets. Again, this is still not to suggest that such outcomes are always attainable, only that, by identifying which layer a policy is designed to address, and understanding which actors have authority in it, more effective results can be achieved.

Thus, U.S. national cybersecurity policy is a textbook example of how the four-layer conceptual model can, and ought to, be applied in understanding complex Internet policy issues.

What does this cybersecurity case study teach us about the four-layer model?

First of all, the U.S. national cybersecurity policy does indeed make explicit attempts at addressing all four Internet layers, and that is an implicit form of recognition of their importance. While the federal government does not state that these are the steps it will take to address this layer, and these another, the fact is that, whether consciously or not, U.S. national cybersecurity policy does address all four Internet layers. Our argument is that this is because, in order for Internet policies to be truly comprehensive, as this one attempts to be, there is an inherent understanding that the infrastructure, technical protocols, software applications, and content all need to be addressed.

Second, policy goals that target certain layers, like influencing the direction of technical standards and protocols, are often more effectively achieved by street-level implementation agents. As illustrated by the NCSD and US-CERT, such agents of implementation are more likely to have the necessary high-level of technical expertise, and therefore be more capable of acting as knowledgeable intermediaries between the public and private sectors as well as in maintaining an active presence in the consortia organizations which direct the technical decision-making of the Internet, in order to meaningfully put into place substantive measures in securing critical cyber assets.

Third, outcomes are attainable where policies target the layer most appropriate for a particular problem. For example, depending on whether the specific type of security threat being addressed is cracktivism or cyberterrorism, organizations' internal policies ought to appropriately call for either technical measures to be deployed (like software filters and network monitors) or infrastructure-based measures (like notifying ISPs and telecom operators). By narrowly targeting a threat occurring at a specific layer, rather than attempting overly broad strokes, more precise defensive measures can be adopted and, thus, security threats can best be mitigated.

Fourth, policies can be designed to target a specific layer with the specific intention of causing cascaded effects at another layer. This crucial principle takes advantage of the fact that the four Internet layers are interdependent and none can reasonably exist and function on their own; rather, they are merely separate parts of what is one coherent larger system. For example, cyberterrorist threats at the Content layer, such as propaganda websites geared toward recruitment, are difficult to shut down due to their transient nature, however, policies can be designed to effectively do so anyway if they target, not the Content itself, but the ISPs and hosting services which provide them their platform. By regulating the service providers, cybersecurity policies can effectively regulate the content that is available on websites, through e-mail, on P2P networks, and more. The Content layer is the most problematic for governments to directly regulate, therefore targeting a different layer is often a better strategic move for generating desired effects upon it.

NOTES

1. Mark Mazzetti and Scott Shane, "Intelligence Official Cites Threat of Cyber-attacks on U.S.," *New York Times* March 12, 2013. Retrieved on March 13, 2013 from http://www.nytimes.com/2013/03/13/us/intelligence-official-warns-congress-that-cyberattacks-pose-threat-to-us.html?hp.

2. John W. Kingdon, *Agendas, Alternatives, and Public Policies* (New York, NY: Longman Press, 2002).

3. Robert O'Harrow Jr., "Cyber Search Engine Shodan Exposes Industrial Control Systems to New Risks," *Washington Post* June 3, 2012. Retrieved on June 5, 2012 from http://www.washingtonpost.com/investigations/cyber-search-engine-exposes-vulnerabilities/2012/06/03/gJQAIK9KCV_story.html.

4. Ellen Nakashima and Joby Warrick, "Stuxnet Was Work of U.S. and Israeli Experts, Officials Say," *Washington Post* June 1, 2012. Retrieved on June 5, 2012 from http://www.washingtonpost.com/world/national-security/stuxnet-was-work-of-us-and-israeli-experts-officials-say/2012/06/01/gJQAlnEy6U_story.html.

5. "CSIS: 20 Critical Security Controls Version 4.1," *SANS*. Retrieved on June 6, 2012 from http://www.sans.org/top20/2007/#s1.

6. "DDOS Attacks in H2 2011," *SecureList*. Retrieved on June 6, 2012 from http://www.securelist.com/en/analysis/204792221/DDoS_attacks_in_H2_2011.

7. Dorothy E. Denning, "Activism, Hacktivism, and Cyberterrorism: The Internet as a Tool for Influencing Public Policy," Paper Sponsored by the Nautilus Institute (RAND 2001). Retrieved October 25, 2005 from http://www.iwar.org.uk/cyberterror/resources/denning.htm.

8. United States Institute of Peace, "Terror on the Internet: Questions and Answers," *USIP.org*. Retrieved on March 11, 2013 from http://www.usip.org/publications-tools/terror-internet/terror-internet-questions-and-answers.

9. Adam Thierer and Clyde Wayne Crews Jr., eds, *Who Rules the Net? Internet Governance and Jurisdiction* (Washington, D.C.: Cato Institute, 2003).

10. U.S. Institute of Peace.

11. "'Miserable Failure' Links to Bush," *BBC News* December 7, 2003. Retrieved on March 6, 2013 from http://news.bbc.co.uk/2/hi/americas/3298443.stm.

12. Robert Domanski, "The Latest Presidential Google Bomb," *The Nerfherder* January 27, 2009. Retrieved on March 6, 2013 from http://thenerfherder.blogspot.com/2009/01/latest-presidential-googlebomb.html.

13. Heather Havebstein, "Blogger Launched 'Google Bomb' at McCain," *ComputerWorld* June 19, 2008. Retrieved on March 6, 2013 from http://www.computerworld.com/s/article/9101218/Blogger_launches_Google_bomb_at_McCain?intsrc=hm_list.

14. Tim Worstall, "The Mitt Romney 'Completely Wrong' Google Bomb," *Forbes* October 14, 2012. Retrieved on March 6, 2013 from http://www.forbes.com/sites/timworstall/2012/10/14/the-mitt-romney-completely-wrong-google-bomb/.

15. Robert Domanski, "The #DontGo Revolution," *The Nerfherder* August 6, 2008. Retrieved on March 6, 2013 from http://thenerfherder.blogspot.com/2008/08/dontgo-revolution.html.

16. Robert Domanski, "Hacktivists vs. Scientology (Again)," *The Nerfherder* February 21, 2008. Retrieved on March 8, 2010 from http://thenerfherder.blogspot.com/2008/02/hacktivists-vs-scientology-again.html.

17. Damien McElroy, "Israeli Battles Rage on Wikipedia," *The Telegraph* May 7, 2008. Retrieved on March 6, 2013 from http://www.telegraph.co.uk/news/1934857/Israeli-battles-rage-on-Wikipedia.html.

18. The Initial Announcement of the protest has been archived by the *Center for Democracy and Technology* and can be found at http://web.archive.org/web/20080117200241/http://www.cdt.org/speech/cda/960203_48hrs_alert.html.

19. Julian Dibbell, "Town Criers for the Net," *Wired.com* May 1996. Retrieved on August 30, 2008 from http://www.wired.com/wired/archive/4.05/scans.html.

20. Scott Shane and Andrew W. Lehren, "Leaked Cables Offer Raw Look at U.S. Diplomacy," *New York Times* November 28, 2010. Retrieved on November 30, 2010 from http://www.nytimes.com/2010/11/29/world/29cables.html?_r=1.

21. The principle that web forums that host user-generated content having limited legal liability stems from Section 230 of the Telecommunications Act of 1996, previously detailed in Chapter 6 on the Content Layer.

22. "What is Hacktivism 2.0?" *TheHacktivist.com* December 2003. Retrieved on August 30, 2008 from http://www.thehacktivist.com/hacktivism.php.

23. "Hippies Declare Web War on WTO," *BBC News* November 30, 1999. Retrieved on October 12, 2005 from http://news.bbc.co.uk/1/hi/uk/543752.stm.

24. Michelle Malkin, "Jihad in Cyberspace," *MichelleMalkin.com* February 20, 2006. Retrieved on February 20, 2006 from http://michellemalkin.com/2006/02/20/jihad-in-cyberspace/.

25. "American Hackers Against Jihad Websites," *KokonutPundit.com* February 20, 2006. Retrieved on February 20, 2006 from http://kokonutpundits.blogspot.com/2006/02/american-hackers-against-jihad.html.

26. Tom Gross, "Israeli Claims to Have Hacked Saddam Off the Net," *London Sunday Telegraph* February 7, 1999.

27. "Israeli-Palestinian Cyber Conflict (IPCC)," White Paper, version 2.0, *iDefense* (Fairfax, VA: iDefense Intelligence Operations, January 3, 2001).

28. James Glave, "Crackers: We Stole Nuke Data," *Wired.com* June 3, 1998. Retrieved on October 10, 2005 from http://www.wired.com/news/technology/0,1282,12717,00.html.

29. Mark Landler and John Markoff, "Digital Fears Emerge After Data Siege in Estonia," *New York Times* May 29, 2007. Retrieved on March 9, 2010 from http://www.nytimes.com/2007/05/29/technology/29estonia.html.

30. Eric Schmitt and Thom Shanker, "U.S. Adapts Cold-War Idea to Fight Terrorists," *New York Times* March 18, 2008. Retrieved on March 9, 2010 from http://www.nytimes.com/2008/03/18/washington/18terror.html.

31. Yochi J. Dreazen, "Military Networks Increasingly Are Under Attack," *Wall Street Journal* March 12, 2008. Retrieved on March 9, 2010 from http://online.wsj.com/article/SB120526061992427783.html.

32. Tania Branigan and Kevin Anderson, "Google Attacks Traced Back to China, says US Internet Security Firm," *The Guardian* January 14, 2010. Retrieved on March 9, 2010 from http://www.guardian.co.uk/technology/2010/jan/14/google-attacks-traced-china-verisign.

33. David E. Sanger, David Barboza, and Nicole Perlroth, "Chinese Army Unit Is Seen as Tied to Hacking Against U.S.," *New York Times* February 18, 2013. Retrieved on March 11, 2013 from http://www.nytimes.com/2013/02/19/technology/chinas-army-is-seen-as-tied-to-hacking-against-us.html?_r=0.

34. John Markoff and Andrew E. Kramer, "U.S. and Russia Differ on a Treaty for Cyberspace," *New York Times* June 27, 2009. Retrieved on March 9, 2010 from http://www.nytimes.com/2009/06/28/world/28cyber.html.

35. National Institute of Standards and Technology, *NIST Incident Handling Guide*. Retrieved on April 13, 2010 from http://csrc.nist.gov/publications/nistpubs/800–61-rev1/SP800–61rev1.pdf.

36. President's Commission on Critical Infrastructure Protection, *Critical Foundations: Protecting America's Infrastructures* (October 1997). Retrieved on November 11, 2008 from http://chnm.gmu.edu/cipdigitalarchive/files/5_CriticalFoundationsPCCIP.pdf.

37. Richard A. Clarke, *Your Government Failed You: Breaking the Cycle of National Security Disasters* (New York, NY: HarperCollins, 2008), 289.

38. *Presidential Decision Directive 63 (PDD-(63): Policy on Critical Infrastructure Protection* (May 22, 1998). Retrieved on July 29, 2008 from http://www.usdoj.gov/criminal/cybercrime/white_pr.htm.

39. Clarke, 292.

40. "Timeline: The U.S. Government and Cybersecurity," *Washington Post* May 16, 2003. Retrieved on June 12, 2008 from http://www.washingtonpost.com/ac2/wp-dyn?pagename=article&node=&contentId=A50606–2002Jun26&notFound=true.

41. Clarke, 292–293.

42. Clarke, 293.

43. "Timeline: The U.S. Government and Cybersecurity."

44. "Timeline: The U.S. Government and Cybersecurity."

45. *National Plan for Information Systems Protection* (January 7, 2000). Retrieved on July 24, 2008 from http://clinton4.nara.gov/media/pdf/npisp-execsummary-000105.pdf.

46. "Timeline: The U.S. Government and Cybersecurity."

47. "Timeline: The U.S. Government and Cybersecurity."

48. The Department of Homeland Security was established by the Homeland Security Act of 2002. It does include implicit references to cybersecurity within its organizational hierarchy, such as the provision for an undersecretary of Science and Technology. See *Homeland Security Act of 2002, Public Law No. 107–296* (November 25, 2002). Retrieved on May 16, 2008 from http://www.dhs.gov/xlibrary/assets/hr_5005_enr.pdf.

49. *Executive Order 132331* (October 16, 2001). Retrieved on May 16, 2008 from http://www.ncs.gov/library/policy_docs/eo_13231.pdf.

50. Critical Infrastructure Protection Board, *53 Questions for Developing the National Strategy to Secure Cyberspace*. Retrieved on May 19, 2008 from http://www.whitehouse.gov/pcipb/53ques.html.

51. *The National Strategy to Secure Cyberspace: Draft for Comments* (September 2002). Retrieved on May 19, 2008 from http://epic.org/security/draftstrategy0902.pdf.

52. "The Cybersecurity Challenge," *Washington Post* December 5, 2003. Retrieved on May 31, 2008 from http://www.washingtonpost.com/wp-dyn/articles/A35977–2003Dec4.html.

53. Robert Vamosi, "We Need a New National Cybersecurity Plan—Now," *CNET* December 10, 2003. Retrieved on May 31, 2008 from http://reviews.cnet.com/4520–3513_7–5112332–1.html.

54. *The National Strategy to Secure Cyberspace: Draft for Comments*, 2.

55. *The National Strategy to Secure Cyberspace*, 2.

56. Anne Larason Schneider and Helen Ingram, *Policy Design for Democracy* (Lawrence, KS: University Press of Kansas, 1997).

57. Davis B. Boborow and John S. Dryzek, *Policy Analysis By Design* (Pittsburgh, PA: University of Pittsburgh Press, 1987).

58. Schneider and Ingram, 68–73.

59. Schneider and Ingram, 69.

60. *The National Strategy to Secure Cyberspace*, 14–15.

61. *The National Strategy to Secure Cyberspace*, 2.

62. *Cyber Security Research and Development Act, Public Law 107–305* (November 2002). Retrieved on July 31, 2008 from http://www7.nationalacademies.org/ocga/laws/PL107_305.asp.

63. U.S. Department of Defense, Memorandum issued by chief information officer John P. Stenbit for Secretaries of the Military Departments, *Subject: Internet Protocol Version 6 (IPv(6)* (June 9, 2003). Retrieved on February 27, 2012 from http://www.defense.gov/news/Jun2003/d20030609nii.pdf.

64. Jeffrey L. Pressman and Aaron Wildavsky, *Implementation*, 3rd ed., expanded (London, England: University of California Press, 1984).

65. D. Van Meter and C. E. Van Horn, "The Policy Implementation Process: A Conceptual Framework," *Administration and Society* 6.4 (1975): 445–488.

66. Paul A. Sabatier and D. A. Mazmanian, "The Implementation of Public Policy: A Framework of Analysis," *Policy Studies Journal* 8 (Special Issue) (1980): 538–560.

67. Eugene Bardach, *The Implementation Game: What Happens After a Bill Becomes Law* (Cambridge, MA: MIT Press, 1977).

68. Benny Hjern, "Implementation Research: The Link Gone Missing," *Journal of Public Policy* 2.3 (1982): 301–308.

69. Michael Lipsky, *Street-Level Bureaucracy: Dilemmas of the Individual in Public Services* (New York, NY: Russell Sage Foundation, 1980).

70. Richard F. Elmore, "Backward-Mapping: Implementation Research and Policy Decisions," *Political Science Quarterly* 94.4 (1980): 601–616.

71. S. M. Barrett and C. Fudge, "Examining the Policy-Action Relationship," *Policy and Action: Essays on the Implementation of Public Policy*, eds. Barrett and Fudge (London, England: Methuen, 1981), 3–34.

72. Michael Hill and Peter Hupe, *Implementing Public Policy: Governance in Theory and in Practice* (London, England: Sage Publications, 2002).

73. Kingdon.

74. Baumgartner and Jones.

75. Lindblom.

76. Scott Charney, "Speech Transcript—Testimony Before the U.S. House of Representatives," *Microsoft News Center* July 24, 2002. Retrieved on March 24, 2010 from http://www.microsoft.com/presspass/exec/charney/07–24testimony.mspx.

77. *The National Cyber Security Partnership*. Retrieved on March 24, 2010 from http://www.cyberpartnership.org/about-overview.html.

78. U.S. Chamber of Commerce, *Commonsense Guide to Cyber Security for Small Businesses* (September 2004). Retrieved on March 24, 2010 from http://www.uschamber.com/sites/default/files/reports/cyberseecurityguide923.pdf.

79. "Cyber Storm: Securing Cyberspace," *U.S. Department of Homeland Security Website*. Retrieved on March 18, 2013 from http://www.dhs.gov/cyber-storm-securing-cyber-space.

80. Dan Verton, "DHS Moves Ahead With Cybersecurity R&D Efforts," *Computerworld* September 15, 2004. Retrieved on May 31, 2008 from http://www.computerworld.com/printthis/2004/0,4814,95942,00.html.

81. *Public Law 107–347: The Federal Information Security Management Act* (2002). Retrieved on May 31, 2008 from http://csrc.nist.gov/drivers/documents/FISMA-final.pdf.

82. "FISMA Implementation Project," *National Institute of Standards and Technology Website—Computer Security Division*. Retrieved on May 31, 2008 from http://csrc.nist.gov/groups/SMA/fisma/index.html.

83. Jason Miller, "Bush Pumps Money Into Cyberdefenses," *FCW.com* November 12, 2007. Retrieved on May 31, 2008 from http://www.fcw.com/online/news/150772–1.html.

84. Michael Rasmussen, Interview with Brian Krebs, "The Cybersecurity Challenge," *Washington Post* December 5, 2003. Retrieved on May 31, 2008 from http://www.security-trends.net/art/secur.html.

85. Robert Vamosi, "We Need a New National Cybersecurity Plan—Now," *CNET.com* December 10, 2003. Retrieved on May 31, 2008 from http://reviews.cnet.com/4520–3513_7–5112332–1.html.

86. Robert Lemos, "Security Guru—Let's Secure the Net," *ZDNET.com* February 19, 2002. Retrieved on May 31, 2008 from http://www.zdnet.com/news/security-guru-lets-secure-the-net/120859.

87. "U.S. Cybersecurity Czar Quits—Again," *CBS News* February 11, 2009. Retrieved on July 29, 2008 from http://www.cbsnews.com/stories/2003/09/15/tech/main573293.shtml.

88. Joel Roberts, "Ex-Bush Aide: Terror War A Bust," *CBS News* February 11, 2009. Retrieved on July 29, 2008 from http://www.cbsnews.com/2100–250_162–560172.html.

89. Ted Bridis, "U.S. Cybersecurity Chief Abruptly Resigns, Cites Frustration," *Seattle Times* October 1, 2004. Retrieved on May 31, 2008 from http://seattletimes.nwsource.com/html/businesstechnology/2002051238_webcybersecurity01.html.

90. President's Information Technology Advisory Committee, *Cyber Security: A Crisis of Prioritization* February 2005. Retrieved on July 23, 2008 from http://www.cyber.st.dhs.gov/docs/PITAC%20Report%202005.pdf.

91. "CSIA Calls for Strategic National Information Assurance Policy," *Government Technology* September 18, 2006. Retrieved on May 31, 2008 from http://www.govtech.com/security/CSIA-Calls-for-Strategic-National-Information.html.

92. *BSA Letter to DHS Secretary Chertoff* July 12, 2006. Retrieved on March 19, 2013 from http://sc-cms.bsa.org/country/News%20and%20Events/News%20Archives/en/2006/en-07122006-letter-chertoff.aspx.

93. Louis Chunovic, "DHS and the $200 Million Cyber Security Mystery," *Government Security News* May 12, 2008. Retrieved on July 23, 2008 from http://www.gsnmagazine.com/cms/features/news-analysis/749.html.

94. John Sterlicchi, "Senators Want Answers on President Bush's Secret Cyber Security Initiative," *Information Security* May 9, 2008. Retrieved on July 23, 2008 from http://www.infosecurity-us.com/news/080509_Bush_cyber_security_initiative.shtml.

95. Chunovic.

96. James Risen and Eric Lichtblau, "Control of Cybersecurity Becomes Divisive Issue," *New York Times* April 16, 2009. Retrieved on June 13, 2012 from http://www.nytimes.com/2009/04/17/us/politics/17cyber.html?_r=1.

97. John O. Brennan, "Time to Protect Against Dangers of Cyberattack," *Washington Post* April 15, 2012. Retrieved on June 13, 2012 from http://www.washingtonpost.com/opinions/time-to-protect-against-dangers-of-cyberattack/2012/04/15/gIQAdJP8JT_story.html.

98. *Cybersecurity Act of 2012, S. 2105 (112th).* Retrieved on June 13, 2012 from http://www.hsgac.senate.gov/download/the-cybersecurity-act-of-2012-s-2105.

99. *Protecting Cyber Space as a National Asset Act, S. 3480.* Retrieved on June 13, 2012 from http://www.hsgac.senate.gov/issues/cybersecurity.

100. *Protecting Cyber Space as a National Asset Act, S. 3480.*

101. *Recommendations of the House Republican Cybersecurity Task Force* (2011). Retrieved on June 13, 2012 from http://thornberry.house.gov/UploadedFiles/CSTF_Final_Recommendations.pdf.

102. *Cyber Intelligence Sharing and Protection Act, H.R. 3523 (112th).* Retrieved on June 14, 2012 from http://thomas.loc.gov/cgi-bin/query/z?c112:H.R.3523.

103. *PRECISE Act, H.R. 3674 (112th).* Retrieved on June 14, 2012 from http://thomas.loc.gov/cgi-bin/query/z?c112:H.R.3674.

104. "DoD Cybersecurity Spending: Where's the Beef?" *Defense Industry Daily* June 14, 2011. Retrieved on June 14, 2012 from http://www.defenseindustrydaily.com/cyber-security-department-defense-spending-06882/.

105. U.S. Department of Defense, *U.S. Cyber Command Fact Sheet* May 25, 2010. Retrieved on June 14, 2012 from http://www.defense.gov/home/features/2010/0410_cybersec/docs/CYberFactSheet%20UPDATED%20replaces%20May%2021%20Fact%20Sheet.pdf.

106. Lewis Page, "Military Set to Lead on U.S. Domestic Cyber-Security," *The Register* May 25, 2011. Retrieved on June 14, 2012 from http://www.theregister.co.uk/2011/05/25/pentagon_lead_us_cyber_security/.

107. "History of Malicious Programs," *SecureList*. Retrieved on February 18, 2010 from http://www.securelist.com/en/threats/detect?chapter=107 .

108. "The Melissa Computer Virus Demonstrates Urgent Need for Stronger Protection Over Systems and Sensitive Data," *Statement of Keith A. Rhodes, GAO Testimony Before the Subcommittee on Technology, Committee on Science, House of Representatives* April 15, 1999. Retrieved on February 18, 2010 from http://www.gao.gov/archive/1999/ai99146t.pdf.

109. *Statement of Keith A. Rhodes*.

110. *US-CERT Website: About Us*. Retrieved on February 12, 2010 from http://www.us-cert.gov/aboutus.html.

111. *US-CERT Website: About Us*.

112. *Public Law 107–296: Homeland Security Act of 2002—Critical Infrastructure Information Act*. Retrieved on February 18, 2010 from http://www.law.umaryland.edu/marshall/crsreports/crsdocuments/RL31762_02282003.pdf.

113. *Protected Critical Infrastructure Information (PCII) Program*. Retrieved on February 18, 2010 from http://www.dhs.gov/files/programs/editorial_0404.shtm.

114. *Multi-State Information Sharing and Analysis Center Website*. Retrieved on February 18, 2010 from http://www.msisac.org/about/.

115. *NIST Website—Computer Security Division*. Retrieved on February 18, 2010 from http://csrc.nist.gov/.

116. *Forum of International Response Security Teams Website*. Retrieved on February 18, 2010 from http://www.first.org/.

117. *U.S. House of Representatives Subcommittee on Emerging Threats, Cybersecurity, and Science and Technology*. Retrieved on February 18, 2010 from http://hsc.house.gov/about/subcommittees.asp?page=2&subcommittee=12&SubSection=0.

118. National Institute for Standards and Technology, *NIST Incident Handling Guide*. Retrieved on February 13, 2010 from http://csrc.nist.gov/publications/nistpubs/800–61/sp800–61.pdf.

119. *NIST Incident Handling Guide*.

120. *National Cyber Security Division Website*. Retrieved on April 13, 2010 from http://www.dhs.gov/xabout/structure/editorial_0839.shtm.

Part IV

CONCLUSION

Chapter 8

Internet Policymaking Moving Forward

When the Web first became popularized in the early 1990s, there was a profusion of libertarian sentiment among pundits and scholars alike in describing the Internet and its political culture, almost bordering on anarchism. In the *Declaration of the Independence of Cyberspace*, John Perry Barlow expressed a common sentiment of the time: "Governments of the Industrial World . . . you have no sovereignty where we gather."[1] One of the Internet's leading architects, David D. Clark, famously proclaimed, "We reject kings, presidents, and voting. We believe in rough consensus and running code."[2] Likewise, MIT's Nicholas Negroponte argued, "It's not that laws aren't relevant, it's that the nation-state is not relevant . . . . The Internet cannot be regulated."[3]

Following this initial wave of ideological enthusiasm, though, came a period of counter-revolutionary thought. A well-founded perception emerged that national governments around the world were aggressively seeking to claim a place at the regulatory table, asserting their territorial jurisdiction in order to transpose their authority to the Internet. This government-centric position is illustrated by scholars Goldsmith and Wu, who claimed, "beneath the fog of modern technology, we have seen the effects of coercive governmental force on local persons, firms, and equipment . . . the United States, China, and Europe are using their coercive powers to establish different versions of what the Internet might be."[4] At the heart of their argument that national governments' authority is vastly underestimated is that the Internet has evolved into a more localized forum, where geography now trumps the "borderless" early Internet due to users wanting content presented in their local language and context and, additionally, advertisers wanting to narrowly target specific audiences that are also more location-specific. This increased localization has led to governments more frequently pressuring or adjudicating local intermediaries—typically private firms or organizations based within their jurisdiction—as a means of controlling Internet content—examples include the French government prohibiting Yahoo from selling Nazi paraphernalia, or the Chinese government restricting what websites can be displayed in Google Search results. The end-result, they argue, is an Internet far more controlled by territorial governments than is often acknowledged.

But as we have seen throughout this project, the truth of Internet governance certainly lies somewhere in between these polar opposite ends of the spectrum. Anarchism doesn't rule the day, nor do governments control everything that occurs in cyberspace. Returning to our original premise, governance of the Internet has indeed emerged, but it is not necessarily governments that are doing most of the governing. Policies are continuously being made that constrain or enable people's behavior on the Internet with intentional effects. Users cannot engage in any activities they desire with an expectation of impunity—their actions are constrained or enabled by policies made at several single controlling points, such as by their ISP, their hosting provider, their network administrator, the large telecommunications providers whose transcontinental networks are being used, as well as, not insignificantly, governments who continue in their attempts to assert their territorial sovereignty.

The anarchist school, including hacker groups and certain elements within the programming community in general, may take issue with these statements arguing that a skilled programmer or hacker could nevertheless circumvent such controlling points and do as they please. In the words of one member of the hacker group Anonymous, "if you know what you're doing, you can travel through the Internet at your will, with no restrictions."[5] They point to recent examples like that of Wikileaks, and its creator Julian Assange, as evidence of how the Internet continues to be "ungovernable" and fosters anarchism—and scholars like Curran and Gibson have given serious consideration to these claims, identifying the anarchical technologies Wikileaks utilizes to foment dissent and the anarchical ethos of its radical politics (although they ultimately determine that Wikileaks is not a "card-carrying doctrinal 'anarchist' organization" but rather is merely in keeping with the contemporary distinction between anarchism per se and the significant influence of anarchist values in oppositional politics).[6]

However, when examined closely, even such statements asserting the boundless abilities of hackers and programmers does not constitute an argument against Internet governance having emerged. First of all, these anarchist-adherents are failing to acknowledge the technical restraints placed upon them by the programming languages and software platforms themselves, which we have explained as implicit controls occurring at the Applications layer. For instance, advanced C++ or .NET programmers might use such languages as highly effective toolkits and be greatly empowered to do much of what they want in cyberspace, however this is only true within the confines of what Microsoft decided C++ as a language would allow. We have examined how the designs of all computer programming languages are the result of explicit decision-making processes, often made by formal institutions and based on political values as much as on technical considerations, and those decisions directly determine the actions of those who deploy them. Thus, the first response to the programmer/hacker critique is that programming languages themselves act as inherent checks on the behavior of programmers.

Second, it must be recognized that there is an additional restraint on programmers—the computing platform. Even if programmers might reluctantly concede to the above assertion, they will then undoubtedly point to alternatives that are not controlled by private commercial firms—examples like nonproprietary open-source languages like PHP, Perl, or Python. If using those languages, wouldn't programmers then be said to be unrestricted in their actions? Not quite. The behavior of the programmer would still be determined by the platform on which the resulting software would be used. The code behind such platforms, whether an operating system like Microsoft Windows, non-OS-dependent platforms like Sun's Java, or various "application programming interfaces" like the Google API, also either constrains or enables the behavior of programmers. The second response to the programmer/hacker critique is that a programmer's code, no matter how independent, must still be written within the confines of rules established by the platform if it wants to achieve a reasonable level of operability.

These are two primary technical restraints refuting anarchist arguments as applied even to those with highly advanced skill-sets, but we cannot ignore significant political restraints as well. Just because laws are circumvented or willfully disobeyed by a relative few does not mean that no governance exists. After all, just because individuals frequently break the speed limit in their cars doesn't mean that governments lack the authority to formulate the rules of the road. They do. Similarly, just because a relatively small number of programmers might mask their IP addresses in order to conduct illegal activities, or a relatively large number of users might engage in file-sharing to download copyrighted music files without paying for them, doesn't mean that there is nobody with authority over their actions. Just ask persons convicted of sharing child pornography online, or anyone who has received a cease-and-desist letter from their ISP for potential copyright infringement. In fact, one could reasonably argue that the very notion that there are measures in place in need of circumvention is proof that a policymaking institution of some kind has such authority in the first place.

Meanwhile, we anticipate another critique to emerge by some scholars who may take issue with our constructed political architecture by oversimplifying it merely as "pluralism." Indeed, there are numerous actors and numerous types of actors all of which, we have argued, have governing authority to some extent. However, as Schneider and Ingram have argued, pluralist theory does not pay sufficient attention to the roles of science and professionalism in shaping policy design choices nor to the pervasive influence of social constructions on policy choices.[7] Such is clearly the case with our political architecture, most clearly evident at the Protocol layer, where the roles of science and professionalism are extremely significant within the governing epistemic community.

Furthermore, our examination of governance *at each layer* provides far more specificity than a simple blanket label of pluralism, or even the Internet-specific label of "accelerated pluralism,"[8] might indicate. For instance, when investigating who governs the Infrastructure layer, we have argued that the Internet's wired infrastructure in the U.S. is governed primarily by the big telecommunication and backbone providers—firms like AT&T, Verizon, Qwest, Level 3, and Sprint Nextel—in conjunction with the FCC who regulates them. This is hardly supportive of an accelerated pluralism argument defined as contributing to the fragmentation of interest-based group politics and leading to less "institutional coherence."[9] Rather, each layer has its own unique political dynamics, and in the case of the wired Infrastructure, we have made the case for an advocacy coalitions framework—focusing on the prominent role of technical expertise, having a preexisting subsystem of policy actors, and incorporating theoretical shifts in value priorities and policy instruments—as best characterizing the current governance dynamic.

In terms of governance theory, some scholars may additionally point to our conclusion of the primacy of private commercial firms as evidence in support of market governance theory. However, doing so would betray, not only the significant role that governments retain, but also, as we have explored, the highly significant contributions of, what Benkler has called, "the economics of non-market social production."[10] The tremendous amounts of Internet activity that have governance effects and are based on, essentially, voluntarism that does not depend on market strategies—from creating open blog publishing platforms like WordPress; to helping develop critical open-source software applications like the Apache Web Server; to contributing time and expertise to the development of next-generation protocols in the IETF—refutes any notion that purely market governance forces are in play.

Thus, as we have demonstrated repeatedly, the Internet is not in a state of anarchy, nor is it controlled by some conspiracy of hidden actors. It is a giant network of privately owned and operated networks, and this decentralized technical architecture leads to a decentralized political architecture as well. Governance, according to our stated definition, occurs primarily in the private sector, but national governments certainly must be recognized to also govern, though to a lesser extent. Foremost, based on how people actually use the Internet, private commercial firms, specifically, are most paramount. Thus, even though every private cyber space is governed separately, in the aggregate whole some actors clearly govern the Internet more than others.

SUMMARY OF FINDINGS

We began this project with three motivating questions: Who governs the Internet? How are they making policies? What are the consequences? We have been especially concerned with distinguishing between those actors who merely have *influence* in policymaking versus those who have a demonstrable governing *authority* to make decisions and set the rules of the environment.

To begin with, let us re-address any skeptics who have doubts that the Internet is governed at all. One need look no further than ICANN and the DNS system for assigning Internet domain names. As we have described, and as other scholars like Milton Mueller have written about at length, the Internet's very functionality on a technical level is completely dependent on this one clearly identifiable institution—ICANN—which is a semipublic international body, and its control over the Internet's core root servers constitute indisputable proof that Internet governance does, in fact, exist.

With that established, by framing the questions of who governs and how are they governing in terms of the Internet's four conceptual layers—the Infrastructure, the Technical Protocols, the Software Applications, and the Content—we were able to formulate substantive answers by constructing a political architecture for the Internet that not only identifies the primary holders of governing authority at each layer, but also provides a descriptive analysis of how those governing actors are actually engaged in policymaking and implementation.

Who governs the Internet's infrastructure? Private telecommunications and cable firms, and the national governments who regulate them. In the United States, for wired communication, this equates specifically to the large telecom firms like AT&T, Verizon, and Qwest, as well as other major backbone players like Level 3 and Sprint Nextel. Meanwhile, the cable industry's authority stems not from ownership of the core infrastructure but from its control over "the last mile" of infrastructure connecting into homes and buildings.

For wireless networks, the spectrum is governed directly by the federal government, while private broadcasters can be said to hold some level of authority only insofar as they have the capacity to make decisions after they have been granted a license to do so by the federal government; and even at that point, their decisions are subject to further governmental oversight and regulatory policy. This is in stark contrast to the governing dynamic of the wired infrastructure, where the private telecoms and cable companies actually own the infrastructure outright. In the context of wireless spectrum, there is a long-established principle that "the public owns the airwaves," thus for nearly a century, the federal government and the FCC, acting as agents of The People, have a demonstrable grip on governing authority.

Who governs the Internet's technical protocols? International consortium groups comprised mainly of scientists, engineers, and academics—foremost among them, the ISOC, its IETF, and the W3C. These international bodies are said to govern because they set all of the rules of the network itself. The developmental processes leading to the adoption of protocols such as TCP/IP, HTTP, 802.11, and IPv6, as well as historical examples like the Internet-OSI standards War, demonstrate how the decisions over which technical protocols to adopt, and how they are to be designed, are, in themselves, an important form of policy that constrain and enable Internet behavior—playing a major role in determining how regulable or non-regulable Web content is going to be. These consortium groups operate largely outside the regulatory grasp of any national government and their policy processes are best characterized by the "rough consensus" principle—an open and transparent process of continuous testing and refinement of proposals that are ultimately measured by responsive models of compliance. Rough consensus agreements within these institutions can usually be placed at around 80–90 percent—"a level high enough to demonstrate strong support, but flexible enough to work in the absence of unanimity." In short, rough consensus is "an informal process in which a proposal must answer to criticisms, but need not be held up if supported by a vast majority of the group."[11]

Who governs the Internet's software applications? We've explored how, because cyberspace is a virtual environment that only exists through software, whoever creates that software is, fundamentally, engaged in a type of policymaking. This is the heart of the "code is law" argument, and we have examined how code constitutes a unique type of policy, one in which the environment itself is designed to deny the user even a *capability* to act in defiance. As a result, virtually any computer programmer who writes web applications' underlying code can be said to govern to some extent, setting the rules for behavior in each of their own private cyber spaces. In practice though, based on usability metrics, it translates into private commercial websites and software firms being the primary governing agents because they set the rules for the vast majority of desktop and web application services that people regularly use and encounter. While fully acknowledging the significant and disruptive influence that individual nonaffiliated programmers like Shawn Fanning, Phil Zimmerman, and Linus Torvalds have demonstrated throughout the Internet's history, ultimately this *influence* has failed to translate into governing *authority*.

Who governs the Internet's content? This is the most politicized and controversial of all four conceptual layers. After examining the three prominent Internet issue areas of the regulation of pornographic material online, the regulation of spam, and the regulation of file-sharing, what they have demonstrated is that governmental policies have often proven to be effective in enabling certain types of Internet content (e.g., Section 230 of the Communications Decency Act), while at other times governmental policies have proven to be inherently limited by the Internet's global scope (i.e.—court rulings on P2P file-sharing) or limited by the technical design of protocols (i.e.—SMTP and the CAN-SPAM Act). Meanwhile, private website operators have exhibited a demonstrable authority to create policies for behavior through their use of code and their TOS Agreements, and private ISPs, likewise, are said to govern via their status as gatekeepers of Internet access.

As a result, the governance of Internet content is best characterized by the theory of "Issue Networks" which accounts for many disparate actors whose webs of influence guide the exercise of power and where participants move in and out of the networks constantly and operate on many levels. Powerful interest groups and knowledgeable individuals alike are often represented, and the "shared-knowledge groups" in the networks are those who are "issue-skilled" regardless of formal professional training. Because of the diversity of policy debates occurring at the Internet's Content layer, issue networks help explain the fluid range of participants involved in each particular policy debate. Furthermore, because the Web consists of so much content, and content of such diverse types, to attempt to define a single, more specific model for policymaking would be to oversimplify the power dynamics that actually occur. Issue networks help to broadly explain the manner in which policies are created in that they account for the enormous range of activities occurring in public policy as it relates to cyberspatial content. For example, the policymaking process regarding the regulation of pornographic material is inevitably going to be different than the process regarding bandwidth caps and Net Neutrality. Different issues involve different debates and different interests, and understanding that the policy networks will not always be homogenous is critical in understanding how policies are being made at the Content layer, in general.

After constructing this political architecture depicting who governs the Internet at each layer, we set out to apply the four-layer model to the case of U.S. national cybersecurity policy, post-9/11. Testing the hypothesis—and commonly held opinion—that the failures of national cybersecurity are the result of a flawed policy design based on public-private partnerships and strictly voluntary measures being adopted, our application of the four-layer model to both the problem stream and the *National Strategy* itself revealed that an implementation process characterized by a lack of funding resources, conflicting roles within the bureaucratic regime, and a high turnover rate among top administrative officials—not policy design—has been the main hindrance to the policy's success to date. Thus, the hypothesis has been refuted.

Despite heated criticism, the voluntary public-private approach prevails, and we have argued that this is because there is no meaningful alternative—a direct consequence of the Internet's decentralized political architecture. With governing authority being so widely dispersed, the federal government does not have the authority on a technical or political level to simply impose its will over actors across all four layers. The Internet is still comprised of millions of independently owned and operated private networks, and each administrator for each private network still sets the rules for what will occur on their own specific piece of cyberspace. As a result, blanket mandates have limited effect, making compliance voluntary by default, and public-private partnerships are an essential policy course to take because the activity that takes place among private actors is really, as we have determined, what is paramount.

This decentralized political architecture is the reason, not only for the reliance on public-private partnerships, but also for the federal government's primary role being relegated to that of a coordinator between numerous private actors. As highlighted in our examination of what actually happens in the face of a widespread cyberattack, private website operators and network administrators are responsible for securing their own private cyber spaces and networks, constituting the nation's "frontline" of cyberdefense, while the federal government, acting through its US-CERT agency within the Homeland Security Department, is charged with being the lead coordinator for information-sharing and response.

By analyzing both the problem stream and policy stream of the cybersecurity issue through the lens of our four layers, this highly complex issue can be more clearly understood in political and technical terms. In terms of the problem stream, the main threat at the Infrastructure layer is the hijacking of core industrial control systems and outright destruction of the infrastructure itself. The main threat at the Applications layer is the infiltration, or cracking, of web application software on both the client and server sides. Finally, the main threat at the Content layer is the defacement of websites or taking them offline completely.

In terms of the policy stream, the four-layer model again proves helpful. Current U.S. national cybersecurity policy, whether codified in the NSSC or implemented through DHS, directly addresses the Infrastructure layer by encouraging the private sector to voluntarily submit vulnerability information to DHS and enables them to do so confidentially. At the Applications layer, the private sector is encouraged to voluntarily deploy technical measures for detection and analysis—typically through software patches and applications and network management tools. Finally, at the Content layer, DHS has set up US-CERT to coordinate information-sharing among the private sector so as to help not only detect cyberattacks but also to mitigate their effects and notify others of the threat.

Overall, the case of U.S. national cybersecurity policy serves to validate our four-layer model and political architecture. Its policy design as well as its implementation illustrates an implicit recognition of the importance of all four Internet layers. Furthermore, it reinforces that the relationship between public and private organizations on the Internet is one of almost total reliance on the private sector in pursuit of public goals—even one as central as national security. Recognizing which actors have a demonstrable authority to govern at each Internet layer is vital to protecting the nation's critical cyber assets, and the coordination-based actions taken by the federal government support the notion that this recognition has, indeed, occurred.

GENERAL CONCLUSIONS & APPLICATION OF THE FOUR-LAYER MODEL

The conclusions of this project are fivefold. First, the four-layer conceptual model serves as a valuable tool for understanding both Internet policies and their underlying political dynamics. Second, and in conjunction with this first principle, the Internet's decentralized technical architecture has led to a decentralized political architecture as well, where numerous governing actors have demonstrable authority over different specific aspects of the Internet—or, in the terminology we have employed, at different layers. It is this political architecture which has occupied much of our focus.

Furthermore, third, and in a prescriptive sense, we have argued that policymakers would be more capable of achieving their policy objectives by narrowly targeting the layer most appropriate to the specific problem they are attempting to address. Fourth, similarly, policy designs ought to, alternatively, target one Internet layer with the express intention of achieving outcomes at a different layer entirely. This is the principle of "cascading effects" and further evidence of the importance of venue selection in strategic policymaking.

Finally, fifth, we strongly affirm the position of Lawrence Lessig and others that technical decisions have inherently political consequences. Code is programmed to embody certain core political values at the expense of others, and therefore the act of creating code has become a very meaningful form of policymaking—and its creators, policymakers.

For one final summary review, let us see how the four-layer model can lead to a better general understanding of complex Internet issues and explore the case of Net Neutrality. Net Neutrality is an issue of tremendous consequence that will directly affect most Americans' online activities for decades to come; yet most people are either completely unaware of it or have little understanding as to what the debate is even about.[12] Net Neutrality has been touted as "The First Amendment of the Internet" and what is specifically at issue are the details of bandwidth capping, network management algorithms, and multitiered-service arrangements. Such issues are hardly for the faint of heart.

But despite Net Neutrality's complexities on a technical level, its politics can be made far more clear to both layman and legislator alike by applying the four-layer model.

We can start by stating that Net Neutrality is an issue that exists at the Infrastructure layer. Immediately, this signals that the debate focuses on how Internet traffic is routed over the infrastructure, and the primary actors with governing authority are the major telecommunications and backbone providers, the cable companies who control the "last mile," and the FCC.

Indeed, the Net Neutrality debate is centered on the actions (or inactions) of the FCC. Their regulatory dilemma: Should broadband providers be legally required to treat all data traversing the network equally, as has been the case since the Internet's inception, or should those providers be free to charge a premium cost to websites that use more of the network's bandwidth for, say, streaming audio and video content?

The pro-neutrality crowd argues that all data must be treated equally in order for the Internet to remain an open marketplace of ideas and innovation. They claim that without Net Neutrality the large telecom companies would create a "toll lane" on the Web, effectively establishing a "tiered Internet" that would grant a *structural* advantage for the most well-capitalized firms, and where entrepreneurs, small businesses, and individuals would all be treated as second-class citizens. Pro-neutrality advocates—comprised of a coalition of large commercial websites like Google and Yahoo, along with smaller websites, civil liberties groups, academics, technologists, and others—believe that FCC guidelines requiring neutrality are necessary in order to guarantee that the Internet continues to exist in its current form.

Meanwhile, the anti-neutrality crowd argues that the government should avoid regulating the Internet and the owners of its infrastructure. These corporate infrastructure owners, they say, will not be blocking access to websites, they will only be making access faster or slower to websites depending on which ones would be willing to pay premium fees. Without neutrality regulations in place, companies like Verizon or Cablevision would be capable of charging a fee to websites like Google (which streams enormous amounts of video through its YouTube site, and thus uses more of the network's bandwidth than sites that are more text-centric), while other service providers like Comcast could almost completely block entire technologies like BitTorrent, that similarly use vast amounts of bandwidth (Comcast has already engaged in such activities).[13] Anti-neutrality advocates argue that telecoms invest billions of dollars into building their network infrastructure, therefore they should be able to make a return on that investment and price their services accordingly.

Regardless of which side's argument one might find more convincing, what is important for our purposes is that, by framing the issue in terms of our four-layer model, the nature of the problem—managing the Internet's physical infrastructure—as well as the governance arrangements surrounding its politics—namely, the roles played by the FCC and the major telecom and backbone firms—are made more easily understandable. The politics of the issue are crystallized as to who actually has the governing authority to make decisions—the FCC and the telecoms—as opposed to who is merely trying to influence those governing actors—private commercial websites like Google and Yahoo, advocacy groups like the EFF, etc.

Thus, our four-layer model leads to the conclusion in the Net Neutrality debate that the anti-neutrality advocates are better positioned in political terms than their pro-neutrality counterparts because the telecoms actually have governing authority to set the policies for their private networks, whereas websites like Google and Yahoo and the other interests mentioned do not. The pro-neutrality crowd can try to influence the decision-making of the telecoms and the FCC, but ultimately, it remains their decision to make.

The four-layer model can not only lead to a better understanding of Internet issues and their politics, it can also serve as a tool for policymakers to better achieve their desired outcomes by designing policies to narrowly target the layer most appropriate to the specific problem they are attempting to address. For example, take the case of Do-Not-Track (DNT).

For years, a number of public interest groups including the World Privacy Forum, the Center for Democracy and Technology, and the EFF lobbied the FTC to create a national DNT registry that would enable Internet users to opt-out of software that allows third-party websites to track their online behaviors—making the assumption that Congress would be amenable toward extending the popular success of its National Do-Not-Call registry which allows people to opt-out of telemarketing phone calls.[14] Indeed, Congressional attempts were subsequently pursued that focused on Content layer measures such as requiring websites to publish what personal information they collect and with whom they share it, prohibiting the collection or sharing of specific types of information including personal medical histories, financial records, or precise geolocation information, and establishing civil penalties of up to $15 million for certain online privacy violations.[15] However, more recently, the strategy of policymakers has shifted away from the Content layer and, instead, toward targeting the Protocol layer. The W3C is currently in the process of standardizing a DNT header field within the HTTP protocol that would universally enable all Internet users to opt-out of tracking through their browsers.[16] The protocol-based argument, as we have previously examined, is that if the ability for users to opt-out of tracking was built into the technology of the network itself, policymakers' desired outcome of enhancing online privacy could be more effectively achieved. This is evidence of how targeting the specific layer most appropriate to a given problem is strategically important to policy designs.

Finally, our conceptual model can also assist policymakers in achieving their desired outcomes by using the principle of "cascading effects" to design policies that target one layer with the express intention of creating effects at another layer entirely. For example, take the case of file-sharing.[17]

As we have previously detailed, the controversial element in the filesharing debate is the illegal dissemination of copyrighted material—comprised mainly of digital music and video. The material in dispute and the legal issues related to copyright infringement all occur at the Content layer. However, efforts by the media industry to clamp down on file-sharing by targeting individual Internet users engaged in such activities have had little to no discernible effect in mitigating the file-sharing phenomenon. Instead, the industry has achieved a modest level of success when strategically targeting the software developers at the Applications layer, bringing lawsuits against the software developers behind Napster, Kazaa, ISOHunt, and others. In the example of file-sharing, intentionally targeting the Applications layer has had considerable cascading effects on behavior at the Content layer.

This is a strategy that can, and should be, replicated. Quite often, directly regulating material at the Content layer is the most problematic, both in technical and legal terms. As a result, those who nevertheless seek to do so ought to design policies that target, not only the ISPs and private website operators at the Content layer, but also, crucially, the software development firms at the Applications layer.

FURTHER

The study of Internet politics and policymaking is still in its nascent stages. Future research ought to explore in greater detail what the cybersecurity case has to say about institution-building and the formation of new bureaucratic regimes. The DHS itself is an enormous and highly important new government apparatus still in its infancy, and the formation and early evolution of its cybersecurity division, specifically, offers researchers a window into something new that can be observed from the ground up. Any path dependencies are still relatively recent or even in flux, which is not frequently the case with bureaucracies in general, and this one is sure to grow in prominence for decades to come.

From a prescriptive point of view, working within the existing public-private framework for cybersecurity, we see an opportunity for the federal government to make greater headway into enhancing national cybersecurity within the private sector by more aggressively using its procurement power to issue, what would essentially be, cybersecurity mandates upon private contractors as a prerequisite for funding, should such "mandates" be deemed desirable. This tactic would have the political benefit of historical precedent, as such procurement powers were a main catalyst used for originally convincing private organizations to connect to the Internet in the early 1970s. The federal government could also apply the cascading effects principle and target acquiring a more prominent influence with increased presence in the major engineering consortium groups at the Protocol layer.

One of our main goals in developing the four-layer conceptual model was to create a new lens for academics and policymakers alike to analyze Internet-related issues that are often highly complex, technical, and frequently misunderstood. Hopefully, future research will be pursued that utilizes this four-layer model in exploring a great range of case studies across the policy spectrum. Breaking down the politics of issues like Net Neutrality, online privacy, or file-sharing—already highly politicized, but also heavily dependent on a large technical decision-making component which is often overlooked by legislators—would be extremely valuable in raising public consciousness and bringing about a better general understanding of Internet issues that are increasingly vital to people's day-to-day lives. What we have presented with the four-layer model is a new framework for analysis, and our hope is that it will prove to be a useful tool for framing a whole range of issues in future research and policymaking.

We have examined how on the Internet, time and again, technical issues have become political, and vice versa. Rules are continuously being produced, and as governments have often been limited in their ability to create those rules, a more complex political architecture of relationships among numerous rule-makers of various different types explains how and why those rules are being made, and, thus, defines how Internet governance is currently constituted. As the Internet continues to become further enmeshed into the political, economic, and cultural fabric of modern society, understanding the calculus of power is as important a task now as ever.

NOTES

1. John Perry Barlow, *Declaration of the Independence of Cyberspace* (February 8, 1996). Retrieved on March 30, 2013 from https://projects.eff.org/~barlow/Declaration-Final.html.

2. David D. Clark, "A Cloudy Crystal Ball—Visions of the Future," Presentation given at the *24th Internet Engineering Task Force* July 16, 1992. Retrieved on March 5, 2011 from http://ietf.org/proceedings/prior29/IETF24.pdf.

3. Nicholas Negroponte, *Being Digital* (New York, NY: Random House, 1995).

4. Goldsmith and Wu, 180, 184.

5. "Interview: Anonymous," *Frontline*, PBS. Retrieved on May 11, 2013 from http://www.pbs.org/wgbh/pages/frontline/shows/hackers/interviews/anon.html.

6. Giorel Curran and Morgan Gibson, "Wikileaks, Anarchism and Technologies of Dissent," *Antipode* 45.2 (March 2013): 294–314.

7. Schneider and Ingram, 66.

8. Bruce Bimber, "The Internet and Political Transformation: Populism, Community, and Accelerated Pluralism," *Polity* 31.1 (August 1998): 133–160.

9. Bimber, 133.

10. Yochai Benkler, *The Wealth of Networks: How Social Production Transforms Markets and Freedom* (New Haven, CT: Yale University Press, 2006).

11. Russell, 48–61.

12. Jon Shingler, "Research Shows that Open Public Debate on Net Neutrality is Critical," *GFK TechTalk* November 11, 2010. Retrieved on March 31, 2013 from http://www.gfktechtalk.com/2010/11/11/research-shows-that-open-public-debate-on-net-neutrality-is-critical.

13. "Comcast Throttles BitTorrent Traffic, Seeding Impossible," *TorrentFreak* August 17, 2007. Retrieved on March 31, 2013 from http://torrentfreak.com/comcast-throttles-bittorrent-traffic-seeding-impossible.

14. "The History of the Do-Not-Track Header," *Slight Paranoia* January 21, 2011. Retrieved on March 31, 2013 from http://paranoia.dubfire.net/2011/01/history-of-do-not-track-header.html.

15. These elements were included in, respectively, the *Do Not Track Me Online Act of 2011* and the *Do Not Track Online Act of 2011*. Retrieved on March 31, 2013 from http://www.gpo.gov/fdsys/pkg/BILLS-112hr654ih/pdf/BILLS-112hr654ih.pdf and http://commerce.senate.gov/public/?a=Files.Serve&File_id=85b45cce-63b3–4241–99f1–0bc57c5c1cff.

16. "Tracking Preference Expression (DNT)," *Tracking Protection Working Group—W3C*. Retrieved on March 31, 2013 from http://www.w3.org/2011/tracking-protection/drafts/tracking-dnt.html.

17. The case of online file-sharing was covered extensively in Chapter 6 on the Content layer.

Glossary

ARPA	Advanced Research Projects Agency, in DoD
AUP	Acceptable Use Policy
BSA	Business Software Alliance
CDA	Communication Decency Act
CERN	European Organization for Nuclear Research
CERT	Computer Emergency Readiness Team
CERT/CC	Computer Emergency Readiness Team Coordination Center
CIA	Central Intelligence Agency
CIAO	Critical Infrastructure Assurance Office
CII	Critical Infrastructure Information Act
CIPA	Children's Internet Protection Act
CIPB	Critical Infrastructure Protection Board
CNCI	Comprehensive National Cybersecurity Iniative
COPA	Children's Online Protection Act
CSIA	Cyber Security Industry Alliance
CYBERCOM	U.S. Cyber Command
DARPA	Defense Advanced Research Projects Agency, in DoD
(D)DoS	(Distributed) Denial of Service attack
DHS	Department of Homeland Security
DNPP	Directorate of National Protection and Programs
DNS	Domain Name System
DoD	Department of Defense
DRM	Digital Rights Management
EDT	Electronic Disturbance Theater
EFF	Electronic Frontier Foundation
EIA	Electronics Industries Alliance
FBI	Federal Bureau of Investigations

FCC	Federal Communications Commission
FedCIRC	Federal Computer Incident Response Center
FISMA	Federal Information Security Management Act
FTC	Federal Trade Commission
GAO	General Accounting Office
IAB	Internet Architecture Board
IAIP	Information Analysis and Infrastructure Protection Directorate
ICANN	Internet Corporation for Assigned Names and Numbers
ICCC	International Conference on Computer Communication
IEEE	Institute for Electrical and Electronics Engineers
IESG	Internet Engineering Steering Group
IETF	Internet Engineering Task Force
IPTO	Information Processing Techniques Office, in DARPA, in DoD
ISA	Internet Security Alliance
ISAC	Information Sharing and Analysis Center
ISO	International Organization for Standardization
ISOC	Internet Society
ISP	Internet Service Provider
ITAA	Information Technology Association of America
ITU	International Telecommunication Union
LAN	Local Area Network
MPAA	Motion Picture Association of America
MS-ISAC	Multi-State Information-Sharing and Analysis Center
NAP	Network Access Point
NCSC	National Cyber Security Center
NCSD	National Cyber Security Division
NCSP	National Cyber Security Partnership
NIAC	National Infrastructure Advisory Council
NII	National Information Infrastructure
NIST	National Institute of Standards and Technology
NSA	National Security Agency
NSC	National Security Council
NSF	National Science Foundation
NSSC	National Strategy to Secure Cyberspace
ODNI	Office of the Director of National Intelligence
OSI	Open Systems Interconnection
OSTP	Office of Science and Technology Policy
P2P	Peer-to-peer
PCCIP	President's Commission on Critical Infrastructure Protection
PCII	Protected Critical Infrastructure Information program

PCIPB	President's Critical Infrastructure Protection Board
PITAC	President's Information Technology and Advisory Committee
PRECISE	Promoting and Enhancing Cybersecurity and Information Sharing Effectiveness Act
PREDICT	Protected Repository for Defense of Infrastructure Against Cyber Threats
RFC	Request for Comments
RIAA	Recording Industry Association of America
SEI	Carnegie Mellon Software Engineering Institute
SMTP	Simple Mail Transfer Protocol
SRI	Stanford Research Institute
TCP/IP	Transmission Control Protocol/Internet Protocol
TOS	Terms of Service
URI	Uniform Resource Identifier
URL	Uniform Resource Locator
US-CERT	U.S. Computer Emergency Readiness Team
W3C	World Wide Web Consortium
WTO	World Trade Organization

Bibliography

Abbate, Janet. "Building the ARPANET: Challenges and Strategies." *Inventing the Internet*. Cambridge, MA: MIT Press, 1999.

Alpert, Jesse and Nissan Hajaj. *We Knew the Web Was Big. . .* July 25, 2008. http://googleblog.blogspot.com/2008/07/we-knew-web-was-big.html.

Atkins, Daniel E., et al. "Revolutionizing Science and Engineering Through Cyberinfrastructure: Report of the National Science Foundation." January 2003.

Aufderheide, Patricia. *Communications Policy and the Public Interest: The Telecommunications Act of 1996*. New York, NY: Guilford Press, 1999.

Bambauer, Derek E. "Solving the Inbox Paradox: An Information-Based Policy Approach to Unsolicited Email Advertising." *Virginia Law Review* 10.5 (Spring 2005).

Barabasi, Albert-Laszlo. *Linked: How Everything is Connected to Everything Else and What it Means for Business, Science, and Everyday Life*. New York: Plume, 2003.

Barbrook, Richard and A. Cameron. *The Californian Ideology*. 1996. June 16, 2006. http://www.hrc.wmin.ac.uk/theory-californianideology.html.

Bardach, Eugene. *The Implementation Game: What Happens After a Bill Becomes Law*. Cambridge, MA: MIT Press, 1977.

Barlow, John Perry. *Decelaration of the Independence of Cyberspace*. 1996. January 5, 2006. http://www.eff.org/~barlow/library.html.

Barrett, S. M. and C. Fudge, eds. "Examining the Policy-Action Relationship." *Policy and Action: Essays on the Implementation of Public Policy*. London, England: Methuen, 1981, 3–34.

Baumgartner, Frank R. and Bryan D. Jones. *Agendas and Instability in American Politics*. Chicago, IL: University of Chicago Press, 1993.

Bell, David. *An Introduction to Cybercultures*. New York, NY: Routledge, 2001.

Benkler, Yochai. "From Consumers to Users: Shifting the Deeper Structures of Regulation." *Federal Communications Law Journal* (2000): 561–563.

———. *The Wealth of Networks: How Social Production Transforms Markets and Freedom*. New Haven, CT: Yale University Press, 2006.

Berners-Lee, Tim and Robert Cailliau. "WorldWideWeb: Proposal for a HyperText Project." November 12, 1990.

Bimber, Bruce. "The Internet and Political Transformation: Populism, Community, and Accelerated Pluralism." *Polity* 31.1 (August 1998): 133–160.

Boborow, Davis B. and John S. Dryzek. *Policy Analysis by Design*. Pittsburgh, PA: University of Pittsburgh Press, 1987.

Bonaccorsi, Andrea and Cristina Rossi. "Comparing Motivations of Individual programmers and Firms to Take Part in the Open Source Movement: From Community to Business." *Knowledge, Technology, & Policy* 18.4 (2006): 40–64.

Bowrey, Kathy. *Law & Internet Cultures*. New York: Cambridge University Press, 2005.

Brafman, Ori and Rod A. Beckstrom. *The Starfish and the Spider: The Unstoppable Power of Leaderless Organizations*. New York, NY: Penguin Group, 2006.

Cerf, Vinton, Yogen Dalal, and Carl Sunshine. "RFC 675: Specification of Internet Transmission Control Program." IETF: Network Working Group, December 1974.

Chadwick, Andrew. *Internet Politics: States, Citizens, and New Communication Technologies*. New York: Oxford University Press, 2006.

Clark, David D. "A Cloudy Crystal Ball: Visions of the Future." Cambridge, MA: Plenary Presentation at the 24th Meeting of the Internet Engineering Task Force, July 13–17, 1992.

Clarke, Richard A. *Your Government Failed You: Breaking the Cycle of National Security Disasters*. New York, NY: Ecco, 2008.

Collins, Richard. *Three Myths of Internet Governance: Making Sense of Networks, Governance and Regulation*. Bristol, UK: Intellect Books, 2009.

Curran, Giorel and Morgan Gibson. "Wikileaks: Anarchism and Technologies of Dissent." *Antipode* 45.2 (March 2013): 294–314.

Dahl, Robert A. *Who Governs? Democracy and Power in an American City*. New Haven, CT: Yale University Press, 1961.

Davis, Richard. *The Web of Politics: The Internet's Impact on the American Political System*. New York: Oxford University Press, 1999.

Denardis, Laura. *Protocol Politics*. Cambridge, MA: MIT Press, 2009.

Drake, William. "The Internet Religious War." *Telecommunications Policy* 17.9 (1993): 643.

Elmore, Richard F. "Backward-Mapping: Implementation Research and Policy Decisions." *Political Science Quarterly* 94.4 (1980): 601–616.

Everard, Jerry. *Virtual States: The Internet and the Boundaries of the Nation-State*. New York: Routledge, 2001.

Foucault, Michel. *Power/Knowledge: Selected Interviews and Other Writings, 1972–1977*. Ed. Colin Gordon. New York, NY: Pantheon, 1980.

Fountain, Jane E. *Building the Virtual State: Information Technology and Institutional Change*. Washington, DC: Brookings Institution Press, 2001.

Franda, Marcus. *Governing the Internet: The Emergence of an International Regime*. Boulder, CO: Lynne Reinner Publishers, 2001.

Frederickson, H. George and Kevin B. Smith. *The Public Administration Theory Primer*. Boulder, CO: Westview Press, 2003.

Galloway, Alexander R. *Protocol: How Control Exists After Decentralization*. Cambridge, MA: MIT Press, 2004.

Goldsmith, Jack and Tim Wu. *Who Controls the Internet? Illusions of a Borderless World*. New York: Oxford University Press, 2006.

Greene, Thomas, Larry James Landweber, and George Strawn. *A Brief History of NSF and the Internet*. National Science Foundation, 2003.

Hajer, Maarten A. and Hendrik Wagenaar, *Deliberative Policy Analysis: Understanding Governance in the Network Society*. Cambridge, UK: Cambridge University Press, 2003.

Hauben, Michael and Ronda Hauben. *Netizens: On the History and Impact of Usenet and the Internet*. Los Alamitos, CA: Wiley-IEEE Computer Society, 1997.

Hill, Carolyn J. and Laurence E. Lynn Jr. "Is Hierarchical Governance in Decline? Evidence from Empirical Research." *Journal of Public Administration Research and Theory* 15.2 (2005): 173–195.

Hill, Kevin and John E. Hughes. *Cyberpolitics: Citizen Activism in the Age of the Internet*. Lanham, MD: Rowman & Littlefield Publishers, Inc., 1998.

Hill, Michael and Peter Hupe. *Implementing Public Policy*. London: Sage Publications, 2002.

Hjern, Benny. "Implemenation Research: The Link Gone Missing." *Journal of Public Policy* 2.3 (1982): 301–308.

Hundt, Reed E. *You Say You Want a Revolution: A Story of Information Age Politics*. New Haven, CT: Yale University Press, 2000.

Jones, Candice, William S. Hesterly, and Stephen P. Borgatti. "A General Theory of Network Governance: Exchange Conditions and Social Mechanisms." *The Academy of Management Review* 22.4 (1997): 911–945.

Jordan, Tim. *Cyberpower: The Culture and Politics of Cyberspace and the Internet*. New York: Routledge, 2000.

Jordan, Tim and Paul A. Taylor. *Hacktivism and Cyberwars: Rebels With A Cause?* New York, NY: Routledge, 2004.

Kamarck, Elaine C. and Joseph S. Nye Jr., *Governance.com: Democracy in the Information Age*. Washington, DC: Brookings Institution Press, 2002.

Katz, Diane and Theodore Bolema. "Crossed Lines: Regulatory Missteps in Telecommunications Policy." 2003.

Keen, Andrew. *The Cult of the Amateur: How Today's Internet is Killing Our Culture*. New York, NY: Random House, 2007.

Kerwin, Cornelius M. and Scott R. Furlong. *Rulemaking: How Government Agencies Write Law and Make Policies*. Washington, DC: Sage Publications, 2011.

Kingdon, John W. *Agendas, Alternatives, and Public Policies*. 2nd ed. New York: Longman, 1995.

Kobrin, Stephen J. "Territoriality and the Governance of Cyberspace." *Journal of International Business Studies* 32.4 (2001): 687–704.

Konieczny, Piotr. "Adhocratic Governance in the Internet Age: The Case of Wikipedia." *Journal of Information Technology and Politics* 7.4 (2010): 263–283.

Krishnamurthy, Sandeep. "Cave or Community? An Empirical Examination of 100 Mature Open Source Projects." *First Monday* (2002).

Leiner, Barry M., et al. *A Brief History of the Internet*. 2003. http://www.internetsociety.org/internet/internet-51/history-internet/brief-history-internet.

———. *A Brief History of the Internet*. 2003.

Lessig, Lawrence. *Code and Other Laws of Cyberspace*. New York: Basic Books, 1999.

———. *Free Culture: The Nature and Future of Creativity*. New York: Penguin Books, 2004.

———. *The Future of Ideas: The Fate of the Commons in a Connected World*. New York: Vintage Books, 2002.

Levmore, Saul and Martha C. Nussbaum, *The Offensive Internet*. Cambridge, MA: Harvard University Press, 2010.

Lindblom, Charles E. "The Science of 'Muddling Through'." *Public Administration Review* 19.2 (1959): 79–88.

Lipsky, Michael. *Street-Level Bureaucracy: Dilemmas of the Individual in Public Services*. New York, NY: Russell Sage Foundation, 1980.

Loader, Brian D., ed. *The Governance of Cyberspace: Politics, Technology, and Global Restructuring*. New York: Routledge, 1997.

Lowi, Theodore J. "Four Systems of Policy, Politics, and Choice." *Public Administration Review* 11 (1972): 298–310.

Lukasik, Stephen J. "Why the ARPANET Was Built." *Annals of the History of Computing, IEEE* 33.3 (2011).

MacLean, Don, ed. *Internet Governance: A Grand Collaboration*. New York: United Nations ICT Task Force, 2004.

McChesney, Robert W. *Telecommunications, Mass Media, and Democracy: The Battle for Control of U.S. Broadcasting, 1928–1935*. New York: Oxford University Press, 1994.

McCubbins, Matthew D. and Thomas Schwartz. "Congressional Oversight Overlooked: Police Patrols versus Fire Alarms." *American Journal of Political Science* 28.1 (February 1984): 165–179.

Mills, C. Wright. *The Power Elite*. New York, NY: Oxford University Press, 1956.

Mintzberg, Henry. *Tracking Strategies: Toward a General Theory*. Oxford, UK: Oxford University Press, 2007.

Morozov, Evgeny. *The Net Delusion: The Dark Side of Internet Freedom*. Philadelphia, PA: Perseus Books, 2011.

Mueller, Milton L. *Ruling the Root: Internet Governance and the Taming of Cyberspace*. Cambridge, MA: MIT Press, 2002.

Murphy, Craig N. "Global Governance: Poorly Done and Poorly Understood." *International Affairs (Royal Institute of International Affairs 1944-)* 76.4 (October 2000): 789–803.

Narten, Thomas. "Internet Routing." *ACM SIGCOMM Computer Communications Review* 19.4 (1989).

"National Strategy to Secure Cyberspace." February 2003. http://www.defense.gov/home/features/2010/0410_cybersec/docs/cyberspace_strategy%5B1%5D.pdf.

Negroponte, Nicholas. *Being Digital*. New York, NY: Random House, 1995.

Nguy, Van N. "Using Architectural Constraints and Game Theory to Regulate International Cyberspace Behavior." *San Diego International Law Journal* 5 (2004): 431.

Okin, J. R. *The Internet Revolution: The Not-for-Dummies Guide to the History, Technology, and Use of the Internet*. Winter Harbor, ME: Ironbound Press, 2005.

Olufs, Dick W. III. *The Making of Telecommunications Policy*. Boulder, CO: Lynne Rienner Publishers, 1999.

O'Neill, Judy E. "The Role of ARPA in the Devvelopment of the ARPANET, 1961–1972." *Annals of the History of Computing, IEEE* 17.4 (1995): 76–81.

Ott, J. Steven and Doug Goodman. "Government Reform or Alternatives to Bureaucracy? Thickening, Tides and the Future of Governing." *Public Administration Review* 58.6 (November–December 1998): 540–545.

Padlipsky, M. A. *Elements of Networking Style*. Englewood Cliffs, NJ: Prentice-Hall, 1985.

Pare, Daniel J. *Internet Governance in Transition: Who is the Master of this Domain?* Lanham, MD: Rowman & Littlefield Publishers, Inc., 2003.

Pelkey, James. "Entrepreneurial Capitalism and Innovation: A History of Computer Communications 1968–1988." 2007. http://www.historyofcomputercommunications.info/Book/4/4.12-ICCC%20Demonstration71–72.html.

Perlman, Brett A. "Pricing the Internet: How to Pay the Toll for the Electronic SuperHighway." *CSIA Discussion Paper 95–01*. Kennedy School of Government, Harvard University, March 1995.

Peters, B. Guy and John Pierre. "Governance Without Government? Rethinking Public Administration." *Journal of Public Administration Research and Theory* 8.2 (April 1998): 223–243.

Pressman, Jeffrey L. and Aaron Wildavsky. *Implementation*. Berkeley, CA: University of California Press, 1984.

Protection, President's Commission on Critical Infrastructure. "Critical Foundations: Protecting America's Infrastructures." October 1997.

Reidenberg, Joel R. "Resolving Conflicting International Data Privacy Rules in Cyberspace." *Stanford Law Review* 52.5 (May 2000): 1315–1371.

Reynolds, Glenn. *An Army of Davids: How Markets and Technology Empower Ordinary People to Beat Big Media, Big Government and Other Goliaths*. Nashville, TN: Nelson Current, 2006.

Ripley, R. B. and G. A. Franklin. *Bureaucracy and Policy: Implementation*. Homewood, IL: Dorsey Press, 1982.

Roberts, Lawrence G. and Barry D. Wessler. "Computer Network Development to Achieve Resource Sharing." *Proceedings of AFIPS*. New York, NY: AFIPS Press, 1970.

Rosenzweig, Roy. "Wizards, Bureaucrats, Warriors, and Hackers: Writing the History of the Internet." *American Historical Review* 103.5 (1998): 1530–1552.

Russell, Andrew L. "'Rough Consensus and Running Code' and the Internet-OSI Standards War." *Annals of the History of Computing, IEEE* (July–September 2006): 48–61.

Russell, Bertrand. *Power: A New Social Analysis*. London, UK: Allen and Unwin, 1938.

Sabatier, Paul A. and D.A. Mazmanian. "The IMplementation of Public Policy: A Framework of Analysis." *Policy Studies Journal* 8 (1980): 538–560.

Sabatier, Paul A., ed. *Theories of the Policy Process*. Boulder, CO: Westview Press, 1999.

Saco, Diana. *Cybering Democracy: Public Space and the Internet*. Minneapolis, MN: University of Minnesota Press, 2002.

Salamon, Lester M. *The Tools of Government: A Guide to the New Governance*. New York, NY: Oxford University Press, 2002.

Schattschneider, E. E. *The Semi-Sovereign People: A Realist's View of Democracy in America*. New York, NY: Holt, Reinhart, and Winston, 1960.

Schneider, Anne L. and Helen Ingram. *Policy Design for Democracy*. Lawrence, KS: University Press of Kansas, 1997.

Shah, Rajiv C. and Jay P. Kesan. "The Privatization of the Internet's Backbone Network." *Journal of Broadcasting and Electronic Media* 51.1 (2007): 93–109.

Shapiro, Andrew. *The Control Revolution: How the Internet is Putting Individuals in Charge and Changing the World We Know*. New York: Public Affairs, 1999.

Solove, Daniel J. *The Digital Person: Technology and Privacy in the Information Age*. New York: New York University Press, 2004.

———. *The Future of Reputation: Gossip, Rumor, and Privacy on the Internet*. New Haven, CT: Yale University Press, 2007.

Spinello, Richard A. *Regulating Cyberspace: The Politics and Technologies of Control*. Westport, CT: Quorum Books, 2002.

Stone, Deborah. *Policy Paradox: The Art of Political Decision-Making*. Revised edition. New York: W.W. Norton & Co., 2002.

Sunstein, Cass R. *Infotopia: How Many Minds Produce Knowledge*. New York, NY: Oxford University Press, 2006.

Surowiecki, James. *The Wisdom of Crowds*. New York, NY: Random House, 2005.

Sutton, Michael F. "Legislating the Tower of Babel: International Restrictions on Internet Content and the Marketplace of Ideas." *Federal Communications Law Journal* 56 (2003): 417.

Tancer, Bill. *Click: What Millions of People Are Doing Online and Why It Matters*. New York, NY: Hyperion, 2008.

Tapscott, Don and Anthony D. Williams. *Wikinomics: How Mass Collaboration Changes Everything*. New York, NY: Penguin Group, 2006.

Taylor, Mark C. *The Moment of Complexity*. Chicago, IL: University of Chicago Press, 2001.

Thierer, Adam and Clyde Wayne Crews Jr., *Who Rules the Net? Internet Governance and Jurisdiction*. Washington, DC: Cato Institute, 2003.

Travis, William. *Networking Basics*. New York, NY: Oxford University Press, 2007.

Tsiavos, Prodromos. *Engineering Policies for the 21st Century: The Open Source Phenomenon and the EU Vision for an Information Society*. London, UK: London School of Economics and Political Science, 2003.

"U.S. Policy Regarding Internet Governance." *The American Journal of International Law* 99.1 (January 2005): 258–259.

Vaidhyanathan, Siva. *The Googlization of Everything (And Why We Should Worry)*. Berkeley, CA: University of California Press, 2011.

Van Meter, D. and C. E. Van Horn. "The Policy Implementation Process: A Conceptual Framework." *Administration and Society* 6.4 (1975): 445–488.

Watson, Richard W. *RFC 372: Notes on a Conversation with Bob Kahn on the ICCC.* IETF: Network Working Group, 12 July 1972.

Weinberger, David. *Everything is Miscellaneous: The Power of the New Digital Disorder.* New York, NY: Henry Holt and Company, 2007.

Winner, Langdon. *Autonomous Technology: Technics Out of Control as a Theme in Political Thought.* Cambridge, MA: MIT Press, 1977.

Wu, Tim. *The Master Switch: The Rise and Fall of Information Empires.* New York, NY: Random House, 2010.

———. "When Code Isn't Law." *Virginia Law Review* 89.4 (2003): 679–751.

Zittrain, Jonathan. *The Future of the Internet and How to Stop It.* New Haven, CT: Yale University Press, 2008.

Zumbansen, Peer. "Rough Consensus and Running Code: A Theory of Transnational Law Making." Berlin, Germany: Paper Presented at the Annual Meeting of The Law and Society Association, July 25, 2007.

Index

About the Author

Robert J. Domanski holds a Ph.D. in Political Science from the City University of New York, where he has taught a wide range of courses in Computer Science, Political Science, and Management Information Systems. His ongoing research is focused on cybersecurity policy, hacktivism, and Big Data.

Lightning Source UK Ltd.
Milton Keynes UK
UKOW01n1121150616

276344UK00009B/381/P

9 781498 512701